THIN CLIENTS

THIN CLIENTS

Dawna Travis Dewire

McGraw-Hill
New York   San Francisco   Washington, D.C.
Auckland   Bogotá   Caracas   Lisbon   London
Madrid   Mexico City   Milan   Montreal   New Delhi
San Juan   Singapore   Sydney   Tokyo   Toronto

Library of Congress Cataloging-in-Publication Data

Dewire, Dawna Travis.
  Thin clients / Dawna Travis Dewire.
      p.   cm.
  Includes index.
  ISBN 0-07-016738-9
    1. Internet (Computer network)   2. Client/server computing.
  3. Intranets (Computer networks)   4. World Wide Web (Information
  retrieval system)   I. Title.
  TK5105.875.I57D48   1998
  004.6—dc21   98-2839
                                                       CIP

McGraw-Hill

A Division of The **McGraw-Hill** Companies

1 2 3 4 5 6 7 8 9 0   FGR/FGR   9 0 3 2 1 0 9 8

ISBN 0-07-016738-9

The sponsoring editor for this book was Michael Sprague, the editing supervisor was Bernard Onken, and the production supervisor was Claire Stanley. It was set in Vendome by Ron Painter of McGraw-Hill's Professional Book Group composition unit.

Printed and bound by Quebecor/Fairfield.

McGraw-Hill books are available at special quantity discounts to use as premiums and sales promotions, or for use in corporate training programs. For more information, please write to the Director of Special Sales, McGraw-Hill, 11 West 19th Street, New York, NY 10011. Or contact your local bookstore.

 This book is printed on recycled, acid-free paper containing a minimum of 50% recycled, de-inked fiber.

To Andy, Travis, and Gregory

CONTENTS

Contents

Contents

Contents

PREFACE

In this split-second world of ours, information is everything. Having the right information for the right people at the right time allows the "right" decision (I tend to think of it as a better-informed decision) to be made. This new technology is about delivering information and services to people—both internal and external to the organization. It provides a new vehicle for customer service and support. It provides a new vehicle for communication between employees and strategic partners. It provides more options for client hardware and server architecture.

This book is intended to educate—to provide some answers, but mostly to raise questions. There is so much information about these newest technologies that it is impossible to know what you don't know. This book is an overview of what you need to know in order to understand what is going on today and where things may be headed. Some parts may be a review for certain readers; for others, this may be totally new information.

The book is divided into five parts. Part 1 discusses what these special kinds of applications are in order to provide a framework for understanding why they are different from the applications we've been developing. Each of the different categories of applications—information sharing, data access, and groupware—is covered in a separate chapter.

The second part of this book deals with the evolution of the Internet and World Wide Web, which focus on one-way sharing of information, into the intranet and extranet applications we are developing today, which are evolving from one-way sharing of information to bidirectional. What has caused this evolution is covered, along with what the risks are. Since Internet, intranet, and extranet architectures share so many common traits, this book uses "i-net" as an umbrella term.

The third part deals with deployment of these new types of applications. Organizations must deal with the hardware side of the architecture, which includes network protocols, network computers and NetPCs, firewalls, and virtual private networks; the infrastructure software, which includes browser and server software and security (message security as well as intruder protection); and application software. The development tools for applications are object-based and require a different approach from 3GL or even 4GL designs. They are based on the use of components.

Once the applications are deployed, they must be managed for responsiveness and reliability. New tools are needed that can handle the connect/disconnect nature of Internet/intranet connections.

The fourth part of this book covers push technology, which is also known as Webcasting. Push technology flips the notion that the client requests information and the server sends it—instead the server just sends information based on a client profile of areas of interest. The client is usually notified that information has arrived, or it may appear as "wallpaper" on the user's desktop. The idea of pushing information is taking a giant step toward acceptance with the integration of push technology into the current versions of the two popular browsers—Microsoft Internet Explorer and Netscape Communicator.

Electronic commerce (E-commerce), the focus of the last section of this book, connects the consumer with a retailer and a supplier with a manufacturer. The consumer uses metaphoric electronic versions of familiar objects such as a shopping cart to pick out items. Retailers use existing applications to process the orders generated through this new interface. Suppliers and manufacturers have been using direct connections to link their internal applications for nearly two decades—electronic data interchange (EDI). These EDI applications have required proprietary software and private networks. The Internet and WWW technology are beginning to change that. A manufacturer can have a Web-based interface into its internal applications, and now even a "small" supplier can use the Internet or a virtual private network link to access those applications. The costs of EDI made its use prohibitive for smaller suppliers, but Web-based interfaces are changing that.

Security continues to be cited as a major stumbling block to the acceptance of E-commerce, but the acceptance of the Secure Electronic Transactions (SET) protocol will overcome that reluctance. In addition, as more and more users use digital certificates, security options increase.

A list of acronyms and definitions can be found following the Glossary.

This is a very exciting technology—more so than client/server computing ever was. I-net technology allows an organization to decide how it can best deliver a process and/or services. It opens an organization to what it might want to do—almost a blue-sky wish list type of approach. People get excited about what they are able to do. There will be no problem getting IT staff to work on these projects—or even end users, for that matter.

This might just be one of those few technologies where it makes sense to develop an application for technology's sake rather than for the "right reason." And isn't that a nice change of pace?

So go ahead, experiment, prototype, draw up some blue-sky wish lists, surf the Net, get some ideas, sign up for push services, get a feeling for what this new era is all about. Have fun along the way, and let me know how you do: DLTDewire@AOL.com.

DAWNA TRAVIS DEWIRE

ACKNOWLEDGMENTS

There are many who continually provided me with encouragement and support over these last few months and never seemed to get tired of hearing, "Sorry, I can't. I have to work on my book." I am very grateful that they never stopped asking.

My colleagues at Babson College have been most supportive. A special thanks to Donna Stoddard, Ted Grossman, and Bob Markus.

And, of course, my family. Travis and Greg took whatever time I could give them, especially toward the end, and are looking forward to more time. None of this would have been possible without Andy's support and willingness to do what he could to help out.

Thank you all for your love and support; it means the world to me.

I-Net
Applications

There are three flavors of applications that use this technology: Internet-based, intranet-based, and extranet-based. They all share the same technology and the same philosophy: the right information to the right people at the right time.

Internet-based applications provide information to the public. They are the marketing-oriented applications that most companies build as their first application of this new technology. People visit the site to gain information about a company's products or services or to get support or technical help, for example. After retrieving the information, the visitor leaves the site to do something with the information.

Intranet-based applications are aimed at the company's employees—the insiders. These applications are "inside the firewall," which refers to the fact that most companies put a firewall between their internal network and their Internet connection port.

Some of these applications may also be information-oriented, such as policies and procedures manuals that are on-line rather than being printed and distributed to all employees. Some are actually application front ends—changing an address for human resources, for example, requesting a room for a meeting and specifying how the room should be set up and ordering the catering, or, in the case of a college, registering for classes.

Extranet applications recognize the fact that there are benefits for allowing strategic partners to access front ends to internal applications, especially legacy applications. An extranet application allows these partners through the firewall based on their user ID and password and allows them to access specific data or initiate specific processes.

Which of these flavors is used depends on what exactly the organization wants to do and how it wants to take advantage of the connectivity and information (and data) retrieval.

CHAPTER 1

What Are I-Net Applications?

The Internet has changed the way companies connect to their customers and trading partners. It is beginning to change the way companies connect to their employees as well. These new applications are called intranets because they use Internet technology over an internal network. If an intranet allows outsiders (trading partners and/or customers) to access it, it becomes an extranet. This book uses *i-net* as an umbrella term to cover all three types of applications—Internet-based, intranet-based, and extranet-based.

1.1 Characteristics of I-Net Applications

Internet-based applications began as interfaces to very static information. Once organizations began to progress up the new technology learning curve, they began to deploy i-net applications with the following characteristics:

- Standards-based
- Dynamic content
- Interactive content
- Powerful GUI components
- Balanced, multitier processing

1.1.1 Standards-based

Key to the use of i-net technology is adherence to standards. As illustrated in Fig. 1-1, these standards allow organizations to pick the best-of-breed for the connections between servers and clients.

1.1.2 Dynamic Content

Larger bandwidth (also called wide pipes) permits rapid delivery of multimedia content. While the speed cannot compare with that of CD-ROMs, companies can use high-resolution graphics and multimedia over their intranets to improve the effectiveness of information exchange.

1.1.3 Interactive Content

Because a fast network is nearly transparent for most client/server applications, intranet applications can provide instant interaction between groups that never had such service—users, workgroups, databases, and other network services. Interactive pages can behave like traditional PC applications, where state information and context govern the application's behavior.

Figure 1-1
Open standards permit choices

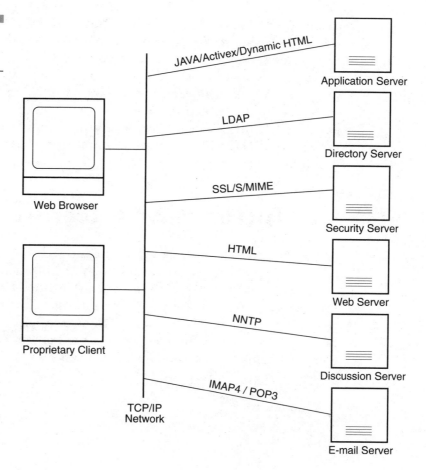

1.1.4   Powerful GUI Components

Hypertext Markup Language (HTML) is well suited for simple screens but is not suited for serious applications, which must provide efficient, information-rich, and usable controls. Organizations use Java applets and ActiveX components to provide these needed features.

1.1.5   Balanced, Multitier Processing

Because the intranet is owned end to end by the corporation, multitier applications with appropriate balancing between client and server processing can be developed. The client-side platform is known, so application and user-interface logic can be deployed there with confidence.

The server side, including back-end databases and other network objects, also is understood and well managed. Information technology (IT) can perform work complementary to the client. With high-performance networks, the coordination between server and client seems instantaneous. Unlike the Internet, where bandwidth limitation dictates that most processing remain on the Web server, intranets can accommodate scalable, network-intensive applications.

1.2   Intranet Architecture

As illustrated in Fig. 1-2, an intranet architecture can be used to connect all the islands of information within an organization, including legacy applications and legacy data. The standard technologies used within the intranet support the notion of "the right data to the right people at the right time." The debate between Intel PCs and Apple Macintosh platforms becomes a nonissue. Protocol conversion between one network and another becomes a nonissue. What becomes the issue is:

- What information do we have?
- Who needs it?
- How do we secure it?

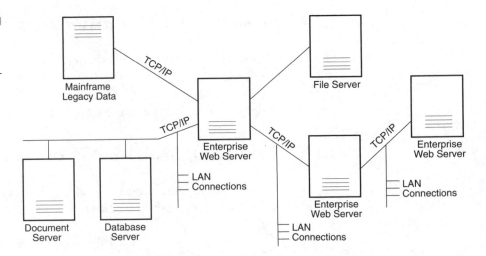

Figure 1-2
An intranet architecture

1.3 Internet vs. I-Net Development

I-net applications should not be developed the same way Internet sites are developed. The only thing they have in common is their technology. Many define an intranet as the same thing as the Internet—it's just on the inside of the firewall. The core components (Web servers, Web browsers, and the network wire between them) form a network-computing platform that is technologically equivalent on both the Internet and an intranet. However, there are several important distinctions between the two.

1.3.1 Intranets Change Business

The Internet is formed by a collection of diverse users and interests (academic, business, and personal). Intranets are developed to meet common business needs and goals, and they are owned by the companies that pay for them.

Investment in intranet technology is increasingly being recognized as the most effective way to automate business processes. The benefits from this investment—cost savings, increased productivity, and competitive advantage—justify exploring this new medium. Investment in the Internet is only as worthwhile as the others that participate on the Internet.

1.3.2 Content-Driven

The content of the Internet, freely available to anyone, is owned in total by no one. This has led to issues of privacy, decency, and security, and often to negative portrayals of the Internet. Intranets are more controlled; the corporation that maintains the intranet has the right to allow only content that meets its business needs and to disallow content that undermines those needs. Consequently, an intranet's content is well targeted, and it results in greater productivity.

1.3.3 Focus on Information

As more companies utilize corporate intranets, the issues facing IT managers will center on creating, distributing, and controlling information. Policy definitions will need to balance the potential for wide-open communication with the need to establish central content and process control. These intranet issues may someday make their way to the Internet, but for now, they live primarily within company walls. How IT managers address them will ultimately lead to the success or failure of an intranet initiative.

Many organizations begin their intranets as a means of Web publishing. Web publishing is very different from print publishing. Because distribution of content is built into the Web model, publishing occurs as soon as content is created and referenced by the Web server. Additionally, browser plug-ins, ActiveX, Java, and browser support of virtually any file format make nearly every file (text, image, audio, etc.) on a network fair game.

As companies invest in new technology such as Java, publishing mechanisms are being developed further to include more sources of live content—applications that perform real work.

1.4   Content

The first step in effectively managing intranet content is understanding what the content is and where it comes from. Intranet content is much more complex to manage than Internet content, chiefly because of the wide variety of information sources available within a company.

1.4.1   Internet Content

The primary difference between publishing on the Internet and publishing on an intranet is in the nature of the content. Content published on the Internet, especially by business-oriented Web sites, has the following characteristics:

- *Aimed at marketing an organization to outsiders—its customers and partners.* This content is usually embellished with attractive graphics, bold pronouncements, and abstract descriptions of the company charter. This content reflects presence on the Internet, but not necessarily substance.

- *Not sensitive.* Few companies provide access to sensitive information on public Web sites.

- *Largely static.* For the most part, content is informational and updated at regular intervals to provide a fresh look for the public. While some companies have simple pages for gathering information from a visitor, few have deployed world-class applications on the Internet.

1.4.2 Intranet Content

An intranet, on the other hand, reveals the inside of a company. How much is viewed depends on how much structure and control is employed. Intranet content has the following characteristics:

- *Deployed across many functional groups and individuals.* There may not be many flashy graphics because intranets focuses on information exchange.

- *Sensitive or confidential.* Because access to an intranet is for employees only, there is a preponderance of sensitive information: future plans, detailed financial data, and competitive information. However, it is important to control access rights even among employees.

- *More dynamic.* The real benefits of an intranet come from using it as a computing platform that performs business functions and not limiting it to providing a useful way to browse static pages.

1.4.3 Sources of Content

The following are potential sources of intranet content:

- *Employees.* Employees are perhaps the largest source of intranet content. The majority of corporate information resides not in relational databases but in user directories, flat files, word-processing documents, spreadsheets, and E-mail messages that have been saved. Intranets tap into this wealth of knowledge.

- *Departments.* Content at the departmental level contains the goals and policies of a group. Much of it is stored in file systems and is not necessarily represented in official corporate databases.

- *Relational databases.* An obvious source of content is relational databases. Corporate data is increasingly finding its way to intranets as distributed Web-based applications replace traditional client/server applications. Because databases hold a company's most valuable information, care should be taken when publishing them on an intranet. As with traditional client/server applications, proper data definition and permission levels are critical.

- *Legacy applications and mainframes.* A company's legacy infrastructure can be leveraged with Web technology. Existing applications, application programming interfaces, and code repositories provide the executable content that can make an intranet an application platform as well as a browsable library.

1.5 Content Delivery

The actual delivery of content through Web applications differs greatly for the Internet and for an intranet. Because intranet sites are built on high-performance networks with well-defined client/server systems, intranet information-delivery techniques and technologies are maturing quickly. Internet sites, by comparison, are slow to mature, held back by lowest-common-denominator solutions.

1.5.1 Intranet-Delivery Possibilities

Two major factors constrain the maturity of Internet applications:

- *Network bandwidth.* Any external Web site must consider the various speeds at which users might connect—many do so at 28.8 kbps or even less. Connections from a corporate backbone, while much faster than personal modems, usually are T1 lines, which are slower than many intranet backbones. These connections are limited further by overloaded Web servers and a relatively slow and outdated Internet infrastructure.

- *Unpredictable computing platform.* We have been led to believe that there is complete compatibility among Web browsers and servers. However, each Web browser and server combination that forms a network-computing platform has its own set of features. As Web product vendors rush to differentiate themselves, one company's feature becomes another company's incompatibility. While Netscape Navigator and Microsoft Internet Explorer both support standard HTML files, many other formats are not supported equally by both. The use of Netscape Plug-ins, as well as differences in Java, JavaScript, and VBScript support among browsers, often leads to "Best viewed with Netscape Navigator" notices on Web sites.

Intranets enjoy some freedom from these constraints.

- *Network performance.* Most companies with serious computing requirements employ networks that far outperform the best of the Internet. Companies commonly use a 100-Mbps fiber-optic backbone, with 10-Mbps LANs serving the desktop. This size bandwidth pipe allows effective use of program logic and user interface objects within Web pages and applications.

■ *Controlled network computing platform.* Because the environment within a corporation is usually well defined, deciding what types of content to deploy is easy. With intranets, companies can survey the domain within corporate walls and decide on the best way to address platform differences. As a result, organizations usually adopt browser and server standards in an effort to minimize interoperability issues and have a more predictable—and flexible—network-computing platform.

1.6 Return on Investment

Intranets are beginning to have to be justified—possibly not on the basis of return on investment (ROI) thresholds or other measurements used in the past for other capital-intensive projects, but justified nevertheless. What needs to be measured is the return from securely selling products (or services) across the Internet and inventing ways to conduct daily business in a secure electronic manner. The measurement also needs to take into account the return from collecting information—collecting valuable data from the customers, not just posting static text on a page for them to read.

Some of the reported ROIs for implementing an intranet are nothing short of astronomical. A recent International Data Corp. (IDC) study pegged ROI at 1000 percent or better—mostly in paper and labor costs. Payback periods ranged from 6 to 12 weeks.

There are many reasons why i-net applications offer such high returns. In many cases, corporations have already built the computing infrastructure needed to run these applications, so the applications don't require a major capital outlay. PCs on user desktops and existing network connections make it possible for users to exchange information. In addition, Web software is inexpensive; browsers are inexpensive or in many cases free, and server software is either free, as with Microsoft's Internet Information Server, or available for a relatively low cost.

High returns are also possible because application development has proved to be much faster than traditional client/server development. New programming tools, such as Sun's Java and Microsoft's ActiveX, as well as a wealth of site-development tools, allow nonprogrammers to build intranet applications in months or even weeks.

The only hidden (actually not very well hidden) catch is increased network traffic. If an employee can examine the value of his or her 401(k) package on a daily basis, it will be checked on a daily basis by most employees.

Cost reduction isn't the only benefit drawing corporations to i-net applications. Other areas where these applications can be used include customer service, maintenance or logistics information, internal document or news distribution, and benefits systems. I-net applications can be used to disseminate incredible amounts of information quickly and easily. Just about any business process that involves passing information from one person to another can be handled more efficiently via an intranet. Information is posted as soon as it's available, instead of waiting for paper distribution—saving time and money.

Another benefit organizations are looking for is quicker and more efficient information retrieval. Now, with electronic storage and search tools, users can zero in on information quickly. They can avoid wading through lengthy paper-based lists, indexes, or tables of contents.

It stands to reason that organizations want to make sure that technology investments either produce more revenue or reduce expenses. Making accurate estimates of exactly how much the return is becomes increasingly difficult. A manufacturer can measure return by examining how many more widgets are produced, or how many fewer employees are needed. Clear cost savings are harder to identify when an improvement means that an employee may spend less time calling coworkers for information or a better decision results from the information that can be accessed.

For this reason, some companies do not complete return on investment studies for their i-net applications. If the i-net application can reduce the critical path for a company's delivery of a service or shorten the sales cycle, then that's a return on the investment. So organizations are evaluating i-net proposals based on such alternatives as

- *Business added value.* The proposal is measured by its support of key company goals and metrics of functional groups rather than by its dollar value.

- *Intangible value.* ROI measures tangible dollars. Intangible values are soft benefits such as improving product quality, staying in the market, and attracting new and retaining existing staff.

- *Net present value.* For some, net present value (NPV) is a subset of ROI in that it is measured in dollars. However, NPV translates tomorrow's returns in today's dollars. It also recognizes that returns from an investment vary from year to year.

1.6.1 ROI Examples

What follows are just a few examples of how organizations are, and are not, evaluating return on investment for i-net applications.

Cadence Design Systems Cadence Design Systems developed OnTrack, which links a home page to information resources and custom applications that map out each phase of the sales process. Employees can gather supporting materials and reference information by simply clicking on a mouse. For ROI, the measurement the company originally wanted to use was how close sales representatives came to meeting their full sales quotas. On closer inspection, the company determined that new sales-persons would have to meet only half of a quota and made the adjustment to their evaluation of ROI. The company continued to refine its measurements, making them more accurate. Ultimately, Cadence determined that the system had a three-year cost of roughly $1.4 million, would pay for itself in a few months, and had a return on investment greater than 1700 percent.

DDB Needham Interactive DDB Needham Interactive is a national advertising agency based in Dallas, with 10,000 employees, stationed in 183 offices throughout 92 countries. The company built an intranet to help its employees win new business. When bidding for new accounts, account executives can now access presentation materials used in similar campaigns. Such information was typically stored in a variety of formats, such as fax or file folders. Even if someone knew to ask for material on particular campaigns, sharing was difficult when employees were in different time zones. Even though the project represented a substantial investment, the company never did a strict ROI analysis. Management felt that there really was no way to measure the potential savings, but knew that the system would quickly pay for itself once an employee won a new account because he or she had access to a coworker's information.

Charles Schwab & Co. Brokerage Charles Schwab & Co. has an intranet called Finance Now that links 600 users in its Electronic Brokerage division to recurring general ledger line items, including salaries, occupancy, communication, advertising, and routine profit and loss data. Previously, each month the division would print 50 copies of the financial information in a 300-page report. Finance Now saves printing costs and training time because users no longer have to be trained in third-party general ledger applications. In addition, analysts used to spend more than 100 hours each month producing the financial report, time that is now spent with customers. Getting information off the Web is also quicker and simpler than paging through the monthly report, and the intranet financials are updated daily so the information is never out-dated.

In 1997, Finance Now's first full year of operation, Charles Schwab expects a return on investment of 600 percent. That figure is expected to rise to 1000 percent in 1998 and 1500 percent by 1999.

1.6.2   Hidden Costs

Keep in mind that these are projections. The initial intranet may be built on existing hardware and inexpensive or free software, but as its use grows, so do its costs.

Intranets have a variety of hidden costs, as illustrated in Fig. 1-3. As the use of the intranet grows, some of the early decisions will prove inadequate. Hardware estimates may be off, growth—in amount of content as well as in number of users—may be underestimated, the impact on the internal network may be underestimated, ease of scalability may be underestimated, etc.

Figure 1-3
Hidden costs for
intranet implementa-
tions

- Hardware will need to be upgraded—while the cost of the hardware itself may seem insignificant, the productivity lost while it is being upgraded is not so insignificant.
- As the Web grows, so does the need for experienced people to support it. These people are hard to find and harder to keep.
- An intranet, especially one that is based on groupware, reengineers business processes to take advantage of the technology. Reengineering is disruptive to an organization—change is always hard.
- Getting the right architecture for an intranet is a matter of trial and error. An intranet that works fine for publishing information won't be suited for collaboration. An architecture for uploading Web pages won't work well for dynamic pages populated from databases in remote locations.
- There are few automated tools for capacity planning, systems management, and monitoring intranet environments. Administration of remote sites has to be automated to be efficient—and it has to be 100 percent reliable.
- Security itself is a big-ticket item. Even the minimum of a firewall and minimum-security controls at the client end can add up to sticker shock.

But as with any new technology, there is also the cost of not trying it at all. It may take a while to figure out how an organization can exploit the new technology. But early adopters that take the chance and find a way have dramatic returns (when all is said and done) and are way ahead of their competitors.

1.7   Internet and Intranet—One Strategy

The Internet and an intranet—companies shouldn't talk about their strategy for one without mentioning the other. Intranet and Internet strategies should complement each other, and perhaps even converge into one. Companies start with pretty Web pages for a presence, then move to a more complex marketing site, from there to an intranet, and possibly to an extranet that serves more complex trading communities.

Intranets need to be noted for more than just raising productivity and improving the bottom line; they need to be closely linked to strategic goals if they are to win acceptance.

Organizations can't change all their legacy environments to browser-based overnight. There needs to be a prioritization done based on the strategic goals of the organization.

In addition to economic considerations, combined i-net strategies and technologies can provide enticing benefits in terms of customer, supplier, and vendor relations. To achieve this, organizations must understand their customers and trading partners—their requirements, their business needs. The new technology should be implemented to provide business solutions and benefits to their customers and trading partners.

At the same time, deployment of these new technologies and business strategies brings about challenges in managing technological change and assuring security, not to mention the challenges presented by the change in the corporate culture.

Federal Express Corp. is an excellent example of turning the convergence of intranet/Internet strategies into a successful model. The award-winning Web site FedEx created lets customers track packages in real time, locate late-night drop boxes, or open shipping accounts. Merchants log onto FedEx BusinessLink and use a Web browser to place package-shipping orders directly. When the order is received on the Federal Express server, the system then electronically transmits information to a server at the merchant's location so that the shipment can be picked up. A customer confirmation number is linked to a Federal Express tracking number, and shipping labels complete with bar codes are printed out near the merchant's site. The system then automatically notifies a Federal Express driver of the pending package pickup. Merchants can ship directly from their desktops to anywhere in the world and not generate one piece of paper to do it.

2

I-Net Applications

If one was to look at all the computerized applications in an organization, two categories that support end users would quickly emerge:

- Access to information (text and data)
- Transaction processing

I-net-based applications are no different. The first Internet-based applications were implemented to allow outsiders to see static information about an organization. The first generation of intranet-based applications (Web broadcasting and Web publishing) has given users access to text information. This capability quickly evolved to being able to access documents. Users then set their sights on data: They wanted business intelligence software—query, reporting, on-line analytic processing, and data mining—from their desktop. And they wanted the Web to be the delivery system. And they wanted it to be portable so that they could work away from the office. A natural outgrowth from all this connectability is Web-based groupware.

Figure 2-1
Original intranet
structure

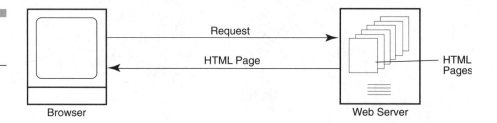

Organizations then look at connecting to trading partners outside their firewall. Customers can order products, review product information, or check on the status of orders. Suppliers and distributors can be linked to internal systems—electronic data interchange. As the technology becomes more secure, organizations are beginning to put transaction-processing applications on their intranets.

The first-level intranets use low-cost browsers and Web servers to share text information that was stored in Hypertext Markup Language (HTML) format, as illustrated in Fig. 2-1. Return on investment is high. As the organization turns to sharing data, the investment increases, but the payback is in the usefulness of the data and better-informed decisions—many organizations feel that an intranet pays for itself with one or two "right" decisions. Transaction processing and groupware require a larger investment, usually in the infrastructure as well as custom coding. Return on investment at this point is even harder to pinpoint—it is often intangible.

2.1   Web Broadcasting

Keeping everyone in an organization informed is a difficult task. Newsletters and memos have been used to spread the word about company news and events. But paper documents take time to create and distribute. They sit in in-boxes until an employee finds time to read them, and by then the content is often no longer relevant. Voice mail and E-mail are faster, but employees have to actively retrieve those messages—still a time-delayed process.

A solution for disseminating information quickly and widely? Intranets. The first thing organizations usually do is create a corporate intranet to give employees access to company information. But many employees are too busy to access the intranet regularly. Intranet broad-

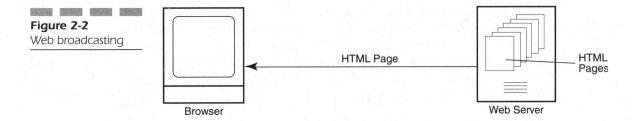

Figure 2-2
Web broadcasting

casting—push technology—is a way to make sure people get the information being distributed. Instead of users asking for this information, it is automatically sent to them, as illustrated in Fig. 2-2. Some organizations use the lure of stock reports as a way to get their employees to rely on their computers for information.

Push technology is discussed in detail in Part 4.

2.2 Web Publishing

Organizations can now use specialized Web servers to automatically convert volumes of documents into HTML as well as to provide search and indexing capabilities to the contents of these documents. This provides an efficient way to make existing marketing, engineering, and training archives available over intranets and public Web sites.

These servers use an on-the-fly approach rather than permanent HTML conversions. By accessing documents composed in Standard Generalized Markup Language (SGML)—the language from which HTML was derived—they make these documents Web-accessible while still retaining their native formats, as illustrated in Fig. 2-3. Publishers can deliver long documents in manageable portions (called chunking) and build indexes between the portions as well as to more material.

A few corporate users already have jumped on the bandwagon. Sun

Figure 2-3
Web publishing

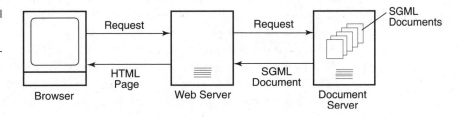

Microsystems has a promotional AnswerBook Online site (already accessible at www.docs.sun.com) which uses a high-end SGML server to deliver documentation for its Solaris products group. Sybase provides the searchable SyBooks-on-the-Web archive (www.sybooks.sybase.com) on more than 40 subjects. In both cases, site servers dynamically generate Web pages from existing repositories.

2.2.1 Manage Documents

Organizations need to manage the documents that are to be published on their intranet. This is not an easy task when you consider that the information that makes up intranet content is usually found in legacy documents—files from non-HTML word processors, databases, and even hard copy.

In most companies, the first candidates for intranet content are the documents of the human resources department—and with good reason. These documents are relatively static, text-based, and generally not time-sensitive. Using them will also build comfort with the technology—like learning Windows and mouse control by playing Solitaire.

For consistency in published documents within an organization, there should be an HTML document template and style guide. This will standardize attributes, formatting, and layouts and also aid Web-page development. In addition, users will be less intimidated if they have a standard Web-page layout to work from—buttons are in the same place on every page, and the layout is the same on every page.

Document-design guidelines that include size as well as content specifications should be in place. Longer documents should be split into smaller pages to improve performance. However, this parsing must be done carefully so as not to lose the reader along the way.

The corporate template and style guide should be tested to ensure that the responsiveness of the client desktop is acceptable and to verify that the pages will display on different browsers.

Graphics are a difficult issue, especially when remote users are involved. One option is to limit the initial display of graphics to thumbnails (a thumbnail is a miniature representation of a page or image) and create links to the full image. An organization may want to establish limits on the size and number of graphics on a page and restrict or ban video and audio clips to conserve space or bandwidth.

Whether legacy documents are published in their native format or as HTML documents, integrity checking is still a must. The organization must ensure that links remain valid and verify that navigational buttons, displays, icons, and backgrounds operate as intended. There should also be a mechanism in place to ensure that documents remain up to date.

The Web server and browser should be able to recognize the application data types being used. Several read-only viewers to consider are Microsoft Word Viewer (www.microsoft.com), Adobe Systems Inc.'s Acrobat Reader and Frame Reader (www.adobe.com), and Corel Corp.'s Envoy (www.corel.com).

Organizations need to be prepared for the overhead required to manage bandwidth and disk space. Not only will an organization need to plan for the overhead of indexes, the content itself will grow exponentially. And so will the number of users and their rate of use.

2.3 Electronic Document Management

Unfortunately, the ease and speed of intranet publishing creates almost as many problems as it solves. Users become overwhelmed by the quantity of information when they can't find the data they need. Webmasters promise that intranet information is current, accurate, and complete, and they then must scramble to fulfill this dream. The result: Companies often still base important business decisions on bad data.

One way to manage the intranet-accessible information is with an electronic document management system (EDMS), which is a collection of complementary technologies that bring together document storage, work flow, and indexing.

Dow Chemical Co. looked to Web-based document management to help organize its growing corporate intranet and automate creation of Web forms. At Dow, document management applications act like a virtual collator. The initial intranet was an instant hit, but the number of documents, both scanned versions of paper documents and forms originally written with HTML, was growing exponentially. At about the same time, Dow was trying to manage the reams of paperwork required by federal chemical industry regulations. Intranet-based document management software was the logical solution.

At Plunkett & Cooney P.C., a 150-attorney firm in Detroit, document management software is allowing both home users and clients to access important documents via the Web. Lawyers working at home can access files at the office. In addition, clients are given password access to files relating to their cases.

2.3.1 Levels of Web-Based Document Management

An electronic document management system can play different roles within an organization. It can

- Make documents viewable (Web-accessible)
- Enable on-line publishing (Web-enabled)
- Facilitate customized communication (Web-exploited)

Each level is fulfilling a different need and comes with different levels of content and process control.

Web-Accessible   A Web-accessible EDMS permits virtually any document, not just those formatted in HTML, to be viewed from a browser. Because all published documents are viewable "as is," the Webmaster does not have to convert and handle documents to allow them to be viewed. This is a huge benefit when the documents are centrally managed and controlled, but enterprise access is required. The downside to this approach is the requirements placed on the browser software. If the browser is opening a non-HTML document, the native application (or a viewer) must be launched in order to view the original formatting. Newer software automatically converts documents to HTML or provides a built-in document viewer that supports a large number of file formats.

Web-Enabled   Web-accessible systems provide read-only access to documents, whereas Web-enabled systems provide a bidirectional working environment. Browser clients can check documents out of a repository, make changes, and check the revised document back into the repository. Web-enabled EDMSs facilitate collaborative document creation over the Web—it becomes a process rather than a publishing mechanism. This could involve inside personnel and outside business partners working together on a contract, customers reviewing and commenting on orders, geographically dispersed team members working to complete a project, or personnel working on a policy and procedures manual.

Web-Exploited   These EDMS applications go to the next level and generate documents that are customized for the viewer. Web pages are constructed on the fly to provide personalized content to the users, who identify themselves as they log onto the system. Customers can receive customized pages that feature their corporate logo, announcements that pertain to products or services they have purchased in the past, or product choices determined previously by contract with this supplier, to name a few. Stock investors can receive reports that focus on their investments only. When employees of a company that has a purchasing relationship with Dell Computers access Dell's home page, they see their company's logo and a list of "approved" products to choose from.

2.3.2 EDMS Components

As illustrated in Fig. 2-4, an EDMS has three major technologies: the repository, the searching-and-indexing technology, and the workflow engine (workflow software). Organizations need to decide which document management features match their business needs and processes.

The Repository The document repository stores, controls, and manages documents. Key repository functions include

- *Library services,* such as controlling access to individual documents, document cataloging, check-in/check-out, and searching for and retrieving documents

- *Version control,* which includes a history of all instances of a document as it changes over time

- *Configuration management,* which is control over the relationships between documents and their component parts, such as a manual and its chapters

Depending on the EDMS vendor, the repository can be simply a database engine or a completely separate application. It can be designed as either a two- or a three-tier system. In a two-tier architecture, the client per-

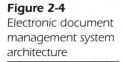

Figure 2-4
Electronic document
management system
architecture

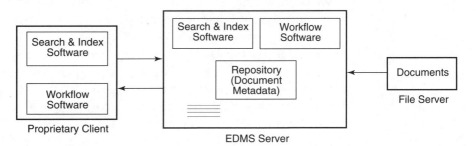

forms more work than in a three-tier environment, where the server is the workhorse.

The database stores all the information about the documents—referred to as metadata—but generally not the documents themselves. The document information typically includes date, author, and title. The database may also store other attributes that the user does not directly provide. Examples include version numbers or, in the case of a document set, pointers that indicate which chapters belong to a particular manual. The database also contains a file pointer, which is the link tying the database and the file system together. The server application controls the file pointer.

Many desktop applications are Object Linking and Embedding (OLE)- or OpenDoc-compliant, which enables them to link content objects from one application to another without requiring any cutting and pasting. EDMS products generally can recognize OLE documents and are intelligent enough to maintain these files and links as relationships in a document repository.

Search Technologies   An EDMS should provide the ability to confine searches to specific attributes, such as date or document type. Such a search returns a handful of entries rather than a list with hundreds of document names, as might be returned from a full-text search.

Repositories can add documents to a full-text index as they are checked in or via a batch job during off-hours to keep the index up to date. The search interface executes attribute searches against a relational DBMS and a word search against the full-text index. The system joins the two results to provide a granular approach to finding needles in haystacks.

EDMS vendors are combining this powerful search capability with Web interfaces to reduce search-engine maintenance from two engines (EDMS and Web site) to one. The full-text engines found on the Web, such as Open Text, Excite, and AltaVista, are also starting to appear in EDMS products.

Workflow Engine   Some electronic document management systems go beyond retrieval of documents—they actually manage the flow of documents. Electronic workflow can eliminate the dead time a document spends in transit between workers and allow people to review a document in parallel instead of serially, which saves time in the sign-off process. Coupled with a repository, workflow applications can provide a full audit history, including review comments. Workflow systems might also notify workers when a new version of a document becomes available. Finally, the workflow engine may drive the conversion process for documents created in one format but distributed (via the Web, for example) in a different format.

Workflow engines typically have two critical integration points in an EDMS: the repository and the E-mail system. The workflow system must have access to the documents and their security, attributes, and other information. Most routing messages from a workflow system go to a proprietary in-box. Rather than having users check yet another in-box, a better solution is to route document-related messages to the E-mail system. However, it is important to note that for security reasons, the document should not be attached to the message.

2.3.3 Managing the Documents

There are currently three models for assembling the material for a Web-based document application.

The Manual Model The manual model, which is used in most Web sites today, provides a way to create documents, convert them to an on-line format such as HTML, and publish them to the Web site. The Webmaster receives any new content and converts these documents to the correct format. The documents are posted to the Web site, and hyperlinks are added to and from the documents.

Unfortunately, the manual process is error-prone and time-consuming. There is also no tracking or other type of document control, because no repository exists in this model. The site becomes as strong as its weakest link—the Webmaster, who is expected to know that a source document changed and ensure that it is converted and placed on-line. This process is informal, and with a site set of any size, it usually breaks down.

The Publishing Model In the publishing model, a repository stores, manages, and controls documents. The publishing step in the process extracts documents from the repository and puts them on the Web site. This is a batch process, and often the workflow engine drives the publishing step, so that updates happen in real time.

However, with this approach, the users can't use all the power of the repository, including attribute searches or security. Another full-text index using the content of the on-line documents must be built. This results in a duplication of effort, and because the on-line documents are detached from the repository, the attributes are not available for searching. Nevertheless, this model can create a high degree of confidence that the intranet information is current.

The Access Model   In the access model, all documents—native and viewable—are stored in the repository. The repository allows Web-based viewing access to the documents. From a user's perspective, the browser seems to be looking at the Web, but it is actually looking into a repository.

By connecting the browser interface to the repository for viewing, the system creates much of the navigation layer on the fly, using the information a document knows about itself. This permits significantly less work in on-line publishing, but it requires that relationships and attributes be kept up-to-date. However, there is only one place to keep the information current.

In addition, users can easily drill down to find the information they need, because the interface shows attribute and hierarchical relationships in the information. The complete search capability of the repository is usually available in this model, as are security schemes to control who can access secure documents.

2.3.4   Evolving Features

Web servers are continually adding more features for document management. However, the most sophisticated management capability in today's Web server is a basic check-in/check-out function. This capability is weak on security and can't handle multiple versions of documents, but for a small HTML-only document set, this alternative is helpful. However, organizations that need to manage native documents, workflow, or the relationships in a compound document should consider a full document management tool.

As with any emerging industry, standards are vital to the successful growth of the market. Standards important to electronic document management systems fall into two categories, interoperability and content-format standards, as seen in Fig. 2-5.

2.4   Web Information Publishing

In today's fast-paced competitive arena, information often provides a major competitive advantage—getting the right information to the right people at the right time in the right format. Many organizations are turning to an intranet technology called Web-to-host publishing, or Web information publishing, to do this.

Web information publishing allows businesses to leverage their investment in existing data warehouses and legacy applications by using the

Figure 2-5
Standards for elec-
tronic document
management systems

Interoperability	
ODMA	Open Document Management API is a set of interfaces between desktop applications (e.g., Microsoft Word) and document management clients. ODMA allows access to a repository without running another application.
DMA	The Document Management Alliance is working toward a set of interfaces between document management servers, so that users can access all repositories in an organization, regardless of the vendor who supplied them.
WfMC	The Workflow Management Coalition is a multipart specification including a common reference model and communications standards for workflow engines, repositories, and mail systems, driving toward the concept of the universal in-box.

Content Format	
SGML	Standard Generalized Markup Language is an ISO standard for creating platform- and application-independent documents. In SGML, a document is divided into three parts: the Document Type Definition or structure, the content, and the formatting. SGML allows documents to know their structure and attributes. It also simplifies information exchange between organizations.
HTML	Hypertext Markup Language is a subset of SGML, but it has evolved into a standard of its own. It is the standard format for Web documents.

TCP/IP transport. It provides a broader base of users with just-in-time information while virtually eliminating the need for manual Web-page creation and updating.

Web information publishing can be extremely cost-effective. The Web is open, easily accessible, and ubiquitous. Using the Web browsers they're already accustomed to, more employees can use business applications and data that have traditionally been off-limits because access was difficult or expensive or both. Support, training, application development, and deployment costs can be substantially reduced in a Web-based environment. Without reengineering the host code, organizations can put a new, user-friendly face on legacy applications.

But while this sounds easy, Web information publishing solutions do have their pitfalls. Careful analysis of the problem and the solution can minimize the hurdles. Some steps organizations need to take are

■ *Consider the access environments.* How many data sources and environments does the Web information publishing solution need to access? Many products are available for only a specific host or server, such as character-based mainframe applications. One solution is products that access mainframe and midrange platforms as well as Open Database

Connectivity (ODBC)–compliant data, data warehouses, custom third-party applications, and E-mail.

- *Design open solutions.* It's important for Web information publishing solutions to be open and compatible with industry-standard servers and browsers.

- *Plan for the future.* A Web information publishing solution should be scalable and extensible. Many solutions work fine in small-scale pilot configurations, but they can't scale up as traffic increases or as new applications and hosts or servers are brought on-line. Other solutions are designed from the start for large-volume, multihost publishing. In addition, software chosen for the system should be flexible enough to accommodate future applications. Some products are easy to implement, but limited in scope and configurability.

- *Determine performance requirements.* Architecture makes a difference in how a Web information publishing solution performs, so it pays to evaluate the technology that the solution employs. Web information publishing solutions built on the new standardized object technology are likely to be faster than those based on a hodgepodge of dated technologies.

- *Decide how much customization will be required.* Solutions should be easily customized at the Web server level without requiring changes to original application code. Web users will therefore see multiple applications and data sources as a single, seamless application. Customization should be integrated, not "pieceware" products that are standalone technologies that must be programmed into larger solutions.

- *Consider how the solution will be supported.* Good administration and security features should be built into the solution, but many products offer very little. The Web information publishing solution should include an administration and configuration component with full access to session and utilization information via the Web browser. Ideally, the support features shouldn't require additional development tools or manual creation/updating of Web pages, should be easy to install and maintain, and should eliminate host downtime.

- *Look for built-in security.* The security features should exploit the security capabilities of the Web server and the host. These capabilities include support for user name and password security and IP address security and the ability to deploy Secure Socket Layer (SSL), Private Communications Technology (PCT), or other Web-based encryption and authentication protocols.

3

Web-based Data Access

One side of the "right information to the right people at the right time" philosophy is data-related. Organizations are looking to intranet technologies for improved and simplified data access. There is beginning to be a convergence of Web technology and data warehouse/data mart technology.

The growth of the Web applications is a result of the open environment created by relational database access protocols. SQL is now the standard access to relational databases. Open Database Connectivity (ODBC) has become the de facto standard in client/server database access.

As thin clients take over from fat clients and the server performs almost all the application interaction and database access, data access requires more robust architectures that distribute functionality across multiple servers.

3.1   Data Access Architectures

Intranet data-access implementations use one of four major architectures: thin client, fat client, server, or distributed.

3.1.1   Thin Clients

Thin clients are best for static reports that require little database access. The thin client communicates via Hypertext Transfer Protocol (HTTP) to a Web server. The Web server's main purpose is to serve documents or files, which may contain very sophisticated graphics that are files in themselves. Common gateway interface (CGI) is used to accept parameters, execute a program, and return the results. For the browser to display results, they must be formatted with Hypertext Markup Language (HTML). CGI-based database gateways can be used to access databases from the Web and display the results of queries. In addition, most database vendors are enabling their databases to work closely with Web servers.

For example, Oracle's Web Application Server acts as an application server that will create a live link between back-end data and a browser-enabled client, allowing users to query and scroll through databases. Oracle, Informix, Sybase, and IBM have created extensions to their core relational databases to perform on-line analytic processing (OLAP) with browsers. In addition, they have created new data types to fit the types of data typically viewed on the Web. Microsoft has ActiveX interfaces for exchanging data between data extraction and query tools.

Web server vendors have also created application programming interfaces (APIs) for server-resident applications, which make applications share a process with the Web server itself. Netscape Server API (NSAPI), Microsoft Internet Services API (ISAPI), and O'Reilly's WebSite API were all designed to improve performance by eliminating the process creation overhead of CGI.

Regardless of how data-access code is generated, the Web development environment using these tools lacks the ability to maintain "state" between actions, a precondition of database access and transactions. HTML and CGI are built to execute independent actions; HTML invokes individual tags or commands, and CGI executes a program and expects a result. But in database access, the software opens the database, prepares an operation, executes it, processes the resultant set, and closes the database—all in sequence. When a typical database connection is initiated, the query state is maintained during processing. This constant "opening" and "closing"—as well as the preparation and execution of the query—creates a great deal of overhead.

With CGI, every time a program is executed, it must manage all phases of database access. A CGI call can't open the database, ready a query, and then wait to execute it the next time the program is run.

3.1.2   Fat Clients

As applications require more functionality and higher graphic and response interfaces for the desktop, the thin client becomes fatter. HTML, although capable of presenting highly graphic materials, does not have the interactive capabilities, such as pull-down menus and list boxes, that today's business users expect. In striving to be portable, HTML sacrificed some of the interactive capability that is proprietary to each operating system.

Plug-ins provide this functionality. Plug-in technologies execute on the desktop while the user is in the browser and provide more interactivity than HTML. In what is known as the "download once, run many times" method, the executable code for the plug-in resides on the server and downloads to the client when necessary. For example, Brio Technologies supports Web-based data queries through interactive plug-ins to its existing product line. InfoSpace has a query and report processing engine written in Java.

However, plugs-in do have drawbacks. For example, they must be downloaded "just-in-time" when needed. Both browsers and clients suffer from their proprietary nature. As a result, it is expected that plug-ins will probably be replaced by Java and ActiveX programs.

3.1.3   Servers

As organizations begin to look at on-line transaction processing (OLTP) and OLAP, they begin to implement a three-tier architecture. Creating a third tier for computing enables certain application tasks to exploit the greater CPU, memory, and I/O capabilities of servers. This architecture uses hardware cost-effectively and eases overall administration.

3.1.4   Distributed Servers

As organizations begin to distribute application components, their architecture becomes multitiered. Application components and databases are distributed across multiple servers, allowing servers to be tuned to specific tasks. Object request brokers (ORBs) become an important component of the architecture. Standards proponents for ORBs are divided into two camps: those supporting Microsoft's Distributed Component Object

Model (DCOM), and those supporting the Object Management Group's Common Object Request Broker Architecture (CORBA). These camps will align with either ActiveX or Java.

I-net technology is not driving object-oriented implementations. Intranets are seen as a low-cost, open, flexible "front-end" to the "back-end" objects. Object-oriented development has been very successful in this arena and is on the rise. However, it requires fundamental changes in the way applications are designed, developed, and deployed.

3.2   Database Queries

Database queries can take several forms, some more complex and resource-intensive than others.

3.2.1   Static Reports

A static report is the simplest type of query and is ideal for the Web. Pre-calculated reports can be put on an intranet much faster and more cheaply than they can be printed and distributed. Training for IT staff usually isn't necessary because most of today's software is Web-aware and can be used to generate Web pages.

3.2.2   Interactive Queries

Interactive queries retrieve a limited set of records that include headers, details, and some lookup tables—exactly what relational databases are designed to do. This type of query is Web-delivered using tools such as CGI and PERL. Today, most relational databases include Web display as an output option for their query languages.

3.2.3   Parameter-Driven Reports

Many reports are parameter-driven. The user provides parameters that are used to filter the returned set—for example, looking up a large number of rows with potential summarization. Unlike the situation with interactive queries, the report's data columns are predetermined. These queries must be written efficiently and effectively because they use substantial resources.

3.2.4 Decision-Support Queries

Decision-support system (DSS)–style queries are free-form and interactive. The user builds the query and then further filters the data to refine the result set. These queries are built much more randomly than other queries.

3.2.5 OLAP Queries

OLAP queries select and manipulate more data than DSS-style queries. OLAP, also known as multidimensional analysis, involves more data aggregation and summarization. Users typically start at the aggregated level of information and drill down to a more detailed level. Often they look at the overall trend and try to understand or substantiate it using the detailed data. Various views of the data are available by pivoting across business dimensions such as product, geography, and customer demographics. Clearly, this type of analysis consumes a great deal of I/O, memory, and CPU cycles.

3.2.6 Data-Mining Queries

Data mining is the most complex kind of query. In this approach, data trends and patterns identify significant information. Data mining involves functions that are much more complex than simple retrieval and algebra. As a result, commercial engines are necessary for data-mining queries and analysis.

3.3 On-line Analytic Processing

On-line analytic processing is not a new technology. As organizations implemented data warehouses which made more data available to users, the users needed to be able to analyze it. OLAP fit the bill.

Most information reporting systems represent data in tabular form. OLAP tools allow users to view the data as having multiple dimensions, which resembles the way the user understands the business, and to then manipulate the multidimensional information. OLAP uses intuitive crosstab reports.

OLAP's principal selling point is flexibility. OLAP offers a drill-down interface that allows users to move through layers of data abstractions until they find an area of interest. By clicking on the data, users can "slice and dice" data to investigate trends, drill down to find the underlying factors that determine a business indicator, and compare values from one part of the business with those from another part. Users look at data as a set of dimensions, which is better suited for analysis than the table rows and columns of relational databases.

For example, a brand manager preparing for an upcoming meeting may need to present a report on the products under her control. The sales data presented by the OLAP tool allow her to look at the data by product, by region, by salesperson, by time, or by any combination or aggregate of these dimensions (products into groups, regions into areas, weeks into months), as illustrated in Fig. 3-1. By slicing through the dimensions, the brand manager can uncover trends and drill down to determine the possible causes of those trends. Trends can be isolated by time, by product, by region, by area, etc., and presented as reports or graphs.

OLAP is a natural fit for server-based processing like that of the Web. Client-based OLAP tools run out of power quickly as business users need more data and analytical capabilities. Shifting database access and analysis to servers expands the utility of these tools. In addition, server-based database access engines can maintain connection and state with the database, improving performance and simplifying access.

OLAP vendors were the first companies to support intranet access. Their products had already moved to a three-tier client/server model, which enabled them to quickly support the server architectural approach of i-nets. Intranet clients decrease the licensing and per-person support costs for deploying OLAP capabilities.

The first generation of Web-based OLAP tools lacked much of the functionality of Windows-based OLAP interfaces. They basically just used HTML interfaces, which offered static reports but no interaction. More sophisticated tools incorporating Java and ActiveX technologies are now being introduced. These tools are including such features as selection, rotation, drill-down, and export; commands that can be embedded into

Figure 3-1
Four dimensions of data

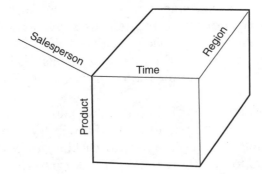

HTML; and the ability to convert existing output or displayed output into Web pages.

One critical issue for OLAP's acceptance on the Web platform is how well the OLAP server will be able to deal with an increased number of users.

3.4   Data Mining

Data mining is part of a larger process called knowledge discovery, which is intended to help organizations find patterns and relationships hidden in their data. Data mining is the step in knowledge discovery that applies specific algorithms to the data to extract possibly useful results. OLAP is often used to verify the patterns and relationships.

Data mining extends decision analysis capabilities provided to business analysts by data warehousing and OLAP—both of which are verification-oriented analysis. Furthermore, data mining allows users to leverage the substantial investments required to build and operate data warehouses.

Traditional database query and report tools are used to describe and extract what is in a database. OLAP is used to analyze the data in the database to determine why certain things are true. The user forms a hypothesis about a relationship and verifies it with a series of queries against the data.

Data mining is used to generate a hypothesis rather than to verify one. For example, an analyst might use data mining to find the risk factors for granting credit. The data-mining tool would analyze the data to discover that people with high debt and low incomes were bad credit risks. But data mining might also discover a pattern that the analyst did not think to try—for example, that education and age are also determinants of risk.

Data-mining tools use statistical and machine-learning methods to search databases for patterns that describe relationships in the data or predict future values or behavior. They look for anomalies and trends. Companies use various data-mining tools and techniques to comb through databases to confirm hypotheses, uncover illuminating correlations, highlight exceptions, or monitor operations more closely.

Data mining is being used to

- Improve both customer acquisition and customer retention. Anything that provides a higher return on marketing expenditures or keeps customers from defecting can pay for itself quickly.

- Reduce fraud. The health care and insurance industries find data mining very promising in reducing fraud.

- Outline inefficiencies or revamp operations.

■ Explore the vast data stores available over the Internet.

Data-mining products should be used only by those who are familiar with the data, the application, and model building. This is a hard set of skills to come by, but it is critical—an analyst may know the business and how to use the tool, but may not be particularly knowledgeable in model building.

To facilitate model building, some products provide a graphical user interface for semiautomatic model building, and others provide a scripting language. Because of important technical decisions in preparing the data and selecting modeling strategies, even with a GUI that simplifies the model building, expertise is needed to create the most effective models.

The key is asking the right questions. But when the right questions are asked, the results can be astounding. Capital One Financial Corp., one of the nation's largest credit card issuers, uses data mining to help identify potential prospects for its 3000 financial products, which include secured, joint, co-branded, and college student cards. When Capital One markets a new product, analysts create a targeted customer list from among the 150 million customers in its 2-terabyte-plus Oracle7–based data warehouse. A strategic benefit of Capital One's data-mining capabilities is fraud detection. Capital One's losses from fraud declined more than 50 percent in 1996.

Other financial companies also are using data mining to improve profitability and reduce risk. Home Savings of America FSB, the nation's largest savings and loan, analyzes mortgage delinquencies, foreclosures, sales activity, and even geological trends over five years to drive risk pricing. KPMG Peat Marwick LLP markets its data-mining capabilities primarily to mortgage banking firms, helping them offer risk-based pricing and reduce fraud. The data-mining analysis can reduce the incidence of fraud because the system looks for anomalies that might indicate fraud, such as a high annual income for a particular area.

The Naval Sea Systems Command aircraft carrier group in Maryland uses a browser-based data-mining application that gathers information on ships, personnel, and finances over the Internet. The browser goes out, and its search agent goes to various databases, extracts related information, and presents it to users in a form they can readily use. Allowing users to relate financial data to information on ships and crews—in nearly real time—makes planning and allocation of resources an easier task.

Data mining can be used to achieve internal as well as external goals. American Express is using a data warehouse and data-mining techniques to reduce unnecessary spending, leverage its global purchasing power, and standardize on equipment and services in its offices worldwide. The system allows American Express to pinpoint, for example, employees who purchase computers or other capital equipment with corporate credit cards meant for travel and entertainment. It eliminates what American Express calls "contract bypass"—purchases from vendors other than those with which the company has negotiated discounts in return for guaranteed purchase levels. By monitoring purchases and vendor performance, American Express can address quality, reliability, and other issues with its vendors.

The Internet, characterized by vast amounts of data, is an emerging arena in which data mining can be used. For example, the Chicago Tribune Co. publishes a variety of services on the Web and on America Online, many of which are focused on classified advertising. The company uses data mining to analyze customers' behavior as they move through its various Web sites in order to make better decisions concerning service and functionality.

3.5 On-line Transaction Processing

As organizations began to implement Web applications, they realized that they could combine an OLTP back end with a Web front end and allow access over TCP/IP. The end result would have the performance of a powerful client/server system without the scalability problems. In addition, the application would adhere to open standards and be accessible via the Internet or the intranet.

A transaction is a collection of actions that make up a single unit of work. Performed together, these actions complete a task. For a transaction to be considered successful, all operations must be performed. If any operation of a transaction cannot be completed, the operations that have taken effect must be undone—rolled back.

In distributed systems, transactions have the following traits, known as the ACID test:

- *Atomicity.* The entire transaction must be either completed or aborted. It cannot be partially completed.

- *Consistency.* The system and its resources go from one steady state to another.

- *Isolation.* The effect of a transaction is not evident to other transactions until the transaction is committed. But any data that a transaction in progress needs is locked to prevent other transactions from changing it.

- *Durability.* The effects of a transaction are permanent and should not be affected by system failures.

For example, consider a simple banking transaction. A customer goes into a bank to transfer $100 from s savings account to a checking account. One part of the transaction is the withdrawal of the money from the savings account; the other part is the deposit into the checking account. If the withdrawal is made but the deposit isn't, the money is in limbo. The customer loses the money from the savings account, but it doesn't show up in the checking account. Both parts of the transaction must occur or neither occurs.

3.5.1 Transaction-Processing Monitors

Organizations have turned to transaction-processing (TP) monitors to manage transactions across their distributed environments. Customer Information Control System (CICS) is a TP monitor, for example. TP monitors became critical components in client/server architectures as organizations tried to manage transactions affecting distributed and replicated data.

A TP monitor is a process manager, transaction manager, and messaging middleware all in one package. The transaction manger guarantees the transaction ACIDity within programs running on the system. The process manager starts server processes, funnels work to them, monitors their execution, and balances their workloads. The message-oriented middleware handles the communications—request/response, queued, and publish-and-subscribe—among the clients and servers.

The requirements for transaction processing on the Web are the same as those required for traditional transaction-processing applications—performance, availability, recoverability, integration with legacy applications, scalability, and security. But the difficulty in fulfilling these requirements is magnified by the Internet's nature.

- The scalability of CGI-based applications is difficult as the number of users increases.
- The Internet has no natural connection to legacy applications.
- The Internet is a stateless environment; it does not maintain the necessary connections to ensure that a transaction is complete.
- Security over the Internet is improving but is still an area of concern.

TP monitors that have been enhanced to support i-net environments address these issues. As illustrated in Fig. 3-2, typically the TP monitor is placed between the database server and the client, thus providing transaction and messaging coordination. By removing business logic from the

Figure 3-2
Web and TP monitor
architecture

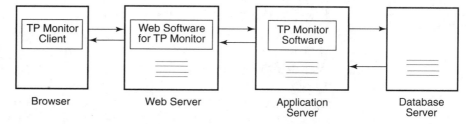

client, TP monitors reduce network traffic and allow databases to improve their transaction-processing performance. TP monitors also allow transactions between heterogeneous databases.

Currently the top TP monitor products in the non-Web world are Tuxedo from BEA Systems Inc., Top End for the Web from NCR Corp., and Transaction Server from IBM. All have been readied for Web interaction.

BEA offers BEA Jolt, which extends Tuxedo to allow it to communicate with Java applets. When an applet is downloaded for a business transaction, Jolt code is downloaded too. The Jolt code knows how to call back to Tuxedo for the services needed. BEA Jolt Java applets work with Tuxedo's Application-to-Transaction Manager Interface APIs to send messages to a Jolt server. This server performs any necessary data marshaling, assembles the request, and interfaces with the Tuxedo TP monitor as a regular transaction. The Jolt server preserves transaction state during execution on requests when multiple interactions occur. Security is handled through Secure Socket Layer (SSL) encryption at the HTML layer and RSA RC4 encryption technology at the link level where the private dialogue takes place.

Top End for the Web allows users to connect to the Web from the Top End TP monitor via a CGI interface or through remote client software written in Java. Whether users choose the CGI or the Java approach depends on the level of interactivity required by the application. Top End for the Web uses distributed Kerberos security. The non-Web product recently added integration with ActiveX, which lets Top End functions appear as objects in applications.

IBM has blended Encina from Transarc Corp. (which was purchased by IBM) with Distributed CICS to create Transaction Server. It is a single transaction-processing engine with a choice of two programming interfaces. With Encina, the DE Light applets package parameters to be executed as a remote procedure call by a gateway residing on a midtier server. The gateway will dynamically build an Encina T-RPC or DCE RPC, which then executes as a normal transaction. DE Light's invocation mechanism is written in Java, so transactions can be invoked from anywhere.

3.5.2 Web Solutions

Microsoft and Netscape Communications have competing platforms intended to support high-volume OLTP on intranets. Currently, corporate developers use Microsoft's Active Platform and Netscape's Open Networking Environment (ONE) for basic lookup and communications functions. But few have entrusted OLTP applications to this new technology. Organizations were hesitant to develop transaction-processing applications using client/server technology, and that hesitancy has carried over to intranets. Few IT managers believe that intranets offer the performance, reliability, and manageability that are needed to process multiple transactions in real time.

There are some success stories, however. Northern Trust Corp. in Chicago offers customers an OLTP application that runs on the company's internal network but is accessed via a World Wide Web browser. The application is based on ONE, Netscape's collection of technologies for building distributed applications. The development team chose Netscape ONE partly for its support of CORBA, since the objects in the application were expected to be a combination of Java and C++.

Turner Broadcasting Sales uses Windows NT to support an order-entry application that the media company built with Microsoft's Active Platform technologies. Microsoft has also promised more object-oriented development in its Active Platform, which is a set of client and server development technologies based on HTML:ActiveX, an open scripting and component architecture. Currently only 200 to 300 employees use the application, so performance problems are rare. The back end of the application uses Microsoft SQL Server database, which sometimes slows during peak usage, but is expected to improve when Microsoft Transaction Server is installed to manage database connections. However, Turner doesn't expect the Transaction Server and SQL Server to support an OLTP application that is currently available on the company's high-traffic Web site with its 13 million hits a day.

3.5.3 Internet Transaction Processing

Some developing OLTP software is being called Internet transaction processing (ITP) and relies heavily on object-oriented programming. ITP environments include WebSpeed from Progress Software Corp. (www.progress.com); Prolifics from Prolifics (www.prolifics.com), a spinoff from Jyacc Inc.; and Studio from NetDynamics Inc (www.netdynamics.com). Major database vendors such as Oracle and Informix have also added Web development tools to their OLTP products.

Even the current TP monitor vendors are starting to integrate more object orientation into their products. BEA purchased Digital's object request broker and expects to release an object TP-middleware product in early 1998. IBM already offers Encina++, which is an object-oriented programming interface for Transaction Server.

4

Web-based Groupware

Groupware software supports a group of users working on a related project who are connected via a network. As the capabilities of technology increase, the concept of groupware continues to evolve. While groupware can be thought of as the document counterpart of database management systems (DBMSs), it is more than multiuser access to the same data. DBMSs typically deal with the highly structured data found in relational databases, whereas groupware tends to deal with less structured data, such as text documents and images.

4.1 Evolution of Groupware

Groupware also provides a mechanism that helps several users coordinate and keep track of an ongoing project. Groupware includes E-mail, group scheduling, and threaded discussions. Currently, if a product is designed to support teamwork, it calls itself groupware. The product to compare groupware products against is Lotus Notes, the first major groupware product, which has continued to maintain a very high installation base. Notes promotes group collaboration—features that allow people to work and communicate with one another electronically—by combining E-mail, document sharing, and data replication and provides an environment for developing applications using these capabilities.

The migration of groupware to the Web hasn't been as speedy as some had expected. Groupware applications threaten some workers who are reluctant to share information with others or resent having software that monitors the flow of work across their desktops. Web-based groupware is even more threatening to those people because Web browsers let an even larger audience participate in the applications. IT organizations cite a lack of Java expertise, which is viewed as key to developing more interactive Web applications, as well as the limitations and late delivery of some products designed to host Web-based groupware.

As traditional groupware vendors hurried to get their products Internet-ready, they interestingly enough turned to proprietary solutions for this industry-standard, open platform. Microsoft uses Outlook 97, a client that runs only on Windows 95 or NT and is designed to work with Microsoft Office. Lotus uses its Notes client. Netscape adds some features that are not available in its Internet-standard offerings. To varying degrees, all the groupware systems today are a blend of proprietary functions and Internet standards.

Moving groupware to intranets is expected to let companies extend the reach of the applications past workers to customers and business partners. The Internet's open protocols also promise the kind of interoperability that will let users book time on one another's schedules even though they are using disparate software packages to track that information. Most organizations are building the infrastructure, getting the technology stable, and allowing the user population to adapt to intranets before they get into groupware. They start by publishing static information on the intranet, which gets employees used to using their computers to get information. Next comes exposing existing data stores to users via Web browsers. For many organizations, the next step is to build server-based interactive applications, such as those that allow workers to participate in workflow processes, over the Web.

4.2 Internet-based Groupware

Internet-based groupware applications—sometimes referred to as intraware—combine E-mail, conferencing, document management, workflow, and group scheduling with the power and flexibility of the World Wide Web.

These groupware products also focus on collaboration, but access is through a Web browser and the software adheres to Internet standards. All of these products offer integrated calendar and scheduling functions, which makes it easier to plan a meeting or collaborate with a team because all of their schedules are visible. These products also allow users to download files from a server and work off-line.

There are currently four major players in this arena:

■ Lotus, with its Lotus Domino server and Lotus Notes client

■ Microsoft, with its Exchange Server and Outlook client

■ Netscape, with its SuiteSpot server and Communicator client

■ Novell, with GroupWise

4.2.1 Lotus Domino Server and Lotus Notes Client

Lotus renamed its Notes product family, so that Notes now refers only to the Notes client, which provides an interface and access to databases, mail, and desktop development tools. The Notes client comes as a full client with application development tools; Notes Desktop, which has all features except the development tools; and Notes Mail, which provides E-mail and access to basic databases.

Domino is a collection of server processes, which include database replicating, mail routing, and indexing. Domino also hosts the Domino Web Server and the Post Office Protocol (POP3) server and Simple Mail Transfer Protocol (SMTP) Message Transfer Agent for E-mail. (SMTP is the TCP/IP protocol that defines the format of an E-mail message.) These servers are all managed from a single server console or a workstation, a remote client, or a Web browser. Domino integrates with Windows NT and supports 16-bit as well as 32-bit environments.

The original Notes strength was its rich-text, document-oriented, flat-file message store with multiprotocol replication and powerful application development tools. In addition, its searching capability could be used to search for words and phrases inside document attachments or across folders and databases using boolean or simple character searches. Domino builds on these strengths by taking Notes views, forms, and documents and converting them to Hypertext Markup Language (HTML) on the fly. Domino allows any user with a Web browser to access these applications, which can be built around any Domino database.

E-mail is central to any groupware product. The Lotus solution allows a user to include rich text, Object Linking and Embedding (OLE) objects, and ActiveX components and Java applets—Lotus calls them Notes Component objects. When a user types a uniform resourse locator (URL) in a message, Notes recognizes and underlines it as a link. URLs can be launched from within messages, from a menu item, or from the Personal Web Navigator. A live Web page can even be embedded into a message, and the page will maintain its links. Mail forms can be completely customized so that interfaces can have the buttons and automation macros that best suit an organization's needs.

New in Notes 4.5 is Calendaring and Scheduling, which allows users to drag and drop to reschedule meetings or tasks on their calendars. Users can check other users' schedules when making a meeting request. A server process called Free Time Lookup checks Notes 4.5 users' schedules and returns free and busy times.

Notes 4.6, released in late 1997, supports POP3/SMTP Internet mail protocols, which allow users to send and receive E-mail from any POP3 account. Notes 4.6 also has a tighter integration with Lotus SmartSuite and Microsoft Office applications. Released at the same time, the full Notes client—renamed Notes Designer for Domino—has improved support for Java and includes Lotus BeanMachine, which is a tool for writing Java applets. These enhancements will make it easier to develop applications for the dual Notes/browser environment.

Domino allows users to access Notes through the Web, which is a great benefit to employees who travel frequently or work from home. Users can send and receive E-mail, view their calendar, and schedule meetings, appointments, or tasks from their Web browser. The Personal Web Navigator can be used for searching the Web and storing favorite sites in a common database. Consequently, when several people are doing research on the Web from different locations, it's easy to keep track of the sites coworkers have already visited.

Domino 4.6, also released in late 1997, supports Internet Message Access Protocol version 4 (IMAP4) and Network News Transfer Protocol (NNTP). It also provides a browser-based administrative tool. In addition, Domino 4.6 includes an updated Hypertext Transfer Protocol (HTTP) server that will serve up Web pages faster.

Lotus has promised some upgrades for mid-1998. The next release of

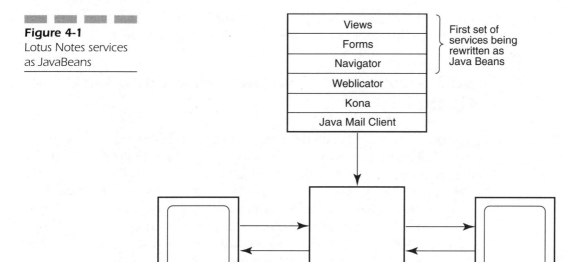

the Notes client, currently referred to as Notes 5.0, will feature an enhanced and simpler user interface that Lotus hopes will reduce the learning curve. In addition, because parts of the Notes client will be written in JavaBeans, as illustrated in Fig. 4-1, they will be accessible from a Domino server through a Notes client or a Web browser. Notes 5.0 will also allow a user to create a "portfolio" of Domino databases for execution of a full-text search. Users will be able to use Word as a message editor instead of the Notes editor. Notes 5.0, which is expected to be released with a new version of the Domino server, currently code-named Maui, will support IMAP4, Lightweight Directory Access Protocol (LDAP), Internet Inter-ORB Protocol (IIOP), and Internet Calendaring Access Protocol, as well as Microsoft's Windows 98 desktop, Channel Definition Format (CDF) technology, and Internet Explorer 4.0.

The new version of Domino, currently referred to as Domino 5.0, will work closely with Windows NT 5.0 (due for release in mid-1998) to enable Domino to synchronize automatically with the NT upgrade's Active Dictionary. Domino 5.0 will support Microsoft's Active Server Page (ASP) technology, Microsoft's Component Object Model (COM) and Distributed COM, and IIOP.

However, the rapid adoption of Microsoft Office 97, which includes Outlook, is giving millions of users a taste of Microsoft Exchange, Notes' biggest rival in the groupware market. Lotus may have good reason to worry. According to many experts, Outlook offers functionality that equals, and in some cases exceeds, that of the Notes client.

4.2.2   Microsoft Exchange Server and Outlook Client

Microsoft Exchange Server depends on Microsoft 32-bit products for operation. Users must be running Windows 95 or Windows NT 4.0 on the desktop to take advantage of the Outlook 97 client.

Exchange Server can run only on an NT server and requires Microsoft's Internet Information Server and Active Server Pages, which are included in NT, for Web integration. Exchange Server can be administered only from an NT server or workstation and not remotely or from a Web browser as with the Lotus solution. Part of the Microsoft BackOffice family of products, Exchange Server provides groupware services for both its proprietary line of clients and its standards-based intranet clients. Based on MAPI (Messaging Application Program Interface), Exchange Server acts as the message and document storage system for various new Microsoft Internet services, such as its NNTP server, LDAP directory, POP3 server, and Internet Mail Services SMTP server.

Outlook offers customizable views that represent each of the groupware functions, such as E-mail, schedule, or contacts. Users can sort, group, and filter any of the items within the views by different criteria. Users also can search the entire contents of Outlook to look for occurrences of a word or phrase. Outlook offers automatic E-mail response, integrated forms creation, and contact management.

Exchange Server stores all the mail folders of a site's users in a single database and all the public folders in another. Domino stores each user's mail as a separate file.

Outlook lets users launch a Web page from within its tools. It recognizes URLs in text, and clicking on a link launches the Web browser. Users can install the ActiveX version of Microsoft Internet Explorer as part of Outlook. Outlook uses Microsoft's VBScript for automating forms and creating custom navigation tools.

Exchange Server allows users to access its mail and public folders with Internet-standard clients. Users can send and receive mail with a POP3 client. Exchange Server translates its proprietary text formatting into HTML, making the text viewable by POP3 clients. The Exchange NNTP server works the same way with Exchange Server's public folders, which are NNTP-enabled by default.

Microsoft's ASP technology is used to create interactive Web applications for Exchange Server. ASPs are HTML pages with ActiveX scripting commands for generating dynamic content. ASPs can call other applications to dynamically generate Web applications.

4.2.3 Netscape SuiteSpot Server and Communicator Internet Client

The SuiteSpot family consists of nine integrated servers, all based around Internet standards or near-standards:

- Mail—SMTP, POP3, IMAP4, and MIME (Multipurpose Internet Mail Extension)
- Web documents—HTTP and HTML
- Discussions—NNTP
- Directory services—LDAP
- Security—Secure Socket Layer (SSL) and X.509

The nine servers are

- *Enterprise Server.* This is Netscape's high-end Web server platform. To integrate the features of the other servers, an HTML page is created and sent to the Enterprise Server for hosting. Enterprise Server also acts as Netscape's file store.
- *Messaging Server.* Previously named Netscape Mail Server, this is an SMTP-, POP3-, and IMAP4-compliant E-mail server. It can be accessed from any standard Internet client via either POP3 or IMAP4.
- *Collabra Server.* On the Internet, the standard for discussion groups is Usenet newsgroups, which are replicated all over the world by NNTP news servers. Collabra Server is a fully compliant NNTP server with two proprietary extensions. One allows users to create new discussion groups; the other allows discussions to be assigned names that are more descriptive than those using the standard dotted-name notation.
- *Calendar Server.* This is the only piece of SuiteSpot that is entirely proprietary. It contains the features one would expect in a calendaring and scheduling product.
- *Directory Server.* This server stores user information and makes it accessible to end users and administrators through LDAP. The Communicator mail client uses LDAP to look up E-mail addresses.

- *Certificate Server.* This server issues and manages digital X.509 certificates that are required for security protocols such as SSL and S/MIME.
- *Media Server.* This server streams audio and video files to clients equipped with the Media Player plug-in.
- *Catalog Server.* This server provides a search engine to query across multiple Web servers.
- *Proxy Server.* This server allows authorized clients to tunnel through firewalls.

Integration across the servers is in administration, security, and the applications development environment. All the SuiteSpot servers are managed through an HTML interface using forms, frames, and JavaScript.

SuiteSpot 3.0 offered these servers as separate components. Netscape announced a restructuring of this product in mid-1997 as SuiteSpot 3.1. This version incorporates a core set of server components which includes the Enterprise Server for Web hosting and additional modules for messaging and collaboration.

The base bundle, called SuiteSpot Standard 3.1, includes Enterprise Server Pro 3.0 with Netcaster, a personal information management tool. The server, which handles content and applications and incorporates the streaming audio features of the Media Server, will ship with an Informix or Oracle database.

The other servers are Messaging Server 3.0, Collabra Server 3.0, and an updated version of Calendar 3.0. The bundle also includes Directory 3.0, which is a core component that is integrated into each individual server.

For more sophisticated enterprises, the SuiteSpot Pro version includes the five base servers of the Standard offering plus Compass 3.0, the next generation of the Catalog Server. The Pro package also includes Proxy Server 2.5 for content replication and filtering, Certificate Server 1.02 for issuing X.509 security certificates, and Mission Control, a set of tools for remotely administering Communicator clients.

The Netscape Communicator client is more a groupware and messaging client than it is a Web browser like Navigator. Communicator integrates with all of SuiteSpot's servers and has client-to-client conferencing. Developers can use JavaScript and Netscape's Dynamic HTML to build applications with database hooks.

4.2.4   Novell GroupWise

For NetWare shops, Novell's GroupWise provides a suite of office applications that integrates E-mail, calendaring, scheduling, and task management as well as support for threaded discussions and document management. Users need TCP/IP on all servers and workstations.

GroupWise uses the Novell Directory Services (NDS) database as its master directory. Because NDS currently runs only on NetWare 4.x, GroupWise requires at least one NetWare 4.x server. NDS is expected to be ported to Windows NT by early 1998.

Three server-based agents handle most of GroupWise's work.

- *Message Transfer Agent* moves messages between domains, post offices, and gateways.

- *Administration Agent* takes information from NDS and puts it into the GroupWise address book.

- *Post Office Agent* handles the actual delivery of messages and other items to the in-box.

Shared folders provide a repository where multiple users can have discussions or store items relevant to the group. Any user can create a shared folder and decide who can read, edit, and delete messages in the folder.

Shared folders are the only items replicated among servers besides NDS information. A replica of a shared folder is created automatically at the post office of each participant. Messages take explicit routes between post offices. If a server along the route is down, GroupWise will hold messages until the connection is reestablished.

2

Evolution

The Internet was once a sleepy little network that tied government agencies and institutions together to support research, offering only E-mail and file transfer. Now it's ubiquitous. Businesses have connections to the Internet, private homes have connections to the Internet (even through their cable companies!), elementary schools have connections to the Internet, upper-level schools have connections to the Internet, kiosks in airports have connections to the Internet.

What turned this sleepy little network into an ever-present force? The browser. The browser lets you view information, jump between pages of information, and follow your train of thought. The first browser didn't set the world on fire (neither, you might recall, did the first version of Windows—remember Version 2.0?), but the second generation did. Now what we talk about is the World Wide Web, which is, to most, the Internet. WWW is the software that makes the network (the Internet) worthwhile today.

How is this possible? Standards. A browser running on any platform can retrieve a Web page and see it without problems because of the use of standards. Messages can go between different types of networks because of the use of standards. Development tools can quickly build applications because of the use of standards.

The Internet and the World Wide Web support the dissemination of information. Organizations are now beginning to use those same standards and technologies to support the internal dissemination of information—intranets. And then they begin to turn their attention to strategic partners who are outside the organization but could benefit from being able to access selected internal information—extranets.

The phenomenon of WWW took all of two years to develop, and it is still evolving every day. It gives organizations options they didn't have before. Most are jumping at the chance.

Stay tuned to an intranet near you for some more remarkable feats.

CHAPTER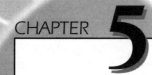

The Internet

The nearly 30-year-old Internet (also referred to as the Net) is a network of connected servers. It was originally used to connect universities and government agencies for research. Eventually E-mail and file transfer became its primary uses.

5.1 The Internet as a Platform

The Internet uses a connection-based pricing model, which means that a user pays a connection fee only to the local node on the network. There are usually no long-distance or per-byte charges for the communication itself.

Organizations started to use the Internet as a way for their geographically dispersed personnel to communicate. That communication quickly went beyond E-mail and file transfers to information dissemination.

Today, the Internet has become ubiquitous. In 1990, there were 93 companies on the Internet. By the end of 1994, there were 21,000. By the end of 1997, there were more than 90,000 companies.

The growth of the Internet has been due in part to its standardization—to permit so many diverse sites to communicate, adherence to standards is a must. Ease of use has been provided via the browser interface software that has found its way to every user's machine.

As growth has spiraled upward, the two problem areas of the Internet, security and data handling, are trying hard to keep up. As the number of users of the Internet expands and more sensitive information (or data) is transmitted over the Internet, security becomes a larger and more complex issue. As Internet technologies become the foundation of mainstream application platforms, data handling has to become easy—and yet secure—for user-oriented applications. Both of these Internet hurdles are discussed throughout this book. Security is discussed in much more detail in Chap. 10, "Infrastructure Requirements," and Chap. 11, "I-Net Security."

5.1.1 Internet2

With its growing commercial popularity, the Internet is running out of capacity, which results in transmission delays. In addition, files travel across the Internet as equals. A video clip that needs to arrive "real time" in a classroom receives the same attention as an E-mail message that may not be read for hours or days.

The university community has joined with government and industry partners to accelerate the next stage of Internet development. The Internet2 project (www.internet2.edu) is working to create a leading edge network capability for the national research community, direct network development efforts to enable applications to fully exploit the capabilities of broadband networks, and improve production Internet services to the broader Internet community.

The project will be conducted in phases over the next three to five years with initial participation from leading research universities, a number of federal agencies, and many of the leading computer and telecommunication companies. Over one hundred universities, each pledging substantial staff resources and financial support, have become members of Internet2. Major corporations such as Ameritech, Cisco Systems, IBM, MCI, Sprint, and Sun Microsystems have also pledged their support.

In 1997, the project's goals were adopted as part of Clinton's White House's Next Generation Internet (NGI) initiative, which expects the second generation of the Internet to communicate in speeds 1000 times faster than today. The federal government will participate in Internet2 through the NGI initiative and related programs.

5.1.2 Internet Standard Technologies

The Internet uses standard technologies to allow communications between two nodes. The Internet E-mail world is in transition from the widely used and fairly basic Post Office Protocol (POP3) to the more sophisticated Internet Access Protocol (IMAP4), which includes features such as adding shared folders and remote access. The other major standards used are Transmission Control Protocol/Internet Protocol (TCP/IP) for the network protocol and HTTP for the document protocol.

5.2 TCP/IP

TCP/IP was designed to allow military research laboratories to communicate if a land war broke out. Independent of any one vendor's hardware, TCP/IP lays out the rules for the transmission of datagrams (transmission units) across a network. It supports end-to-end acknowledgment between the source and destination of the message, even if they reside on separate networks. TCP/IP is the transportation protocol used by the Internet.

As illustrated in Fig. 5-1, TCP/IP is a four-layer architecture that is built on a physical network interface, and that specifies conventions for communications and network interconnection and traffic routing. It allows networks to communicate by assigning unique addresses to each network.

Figure 5-1
TCP/IP's architecture

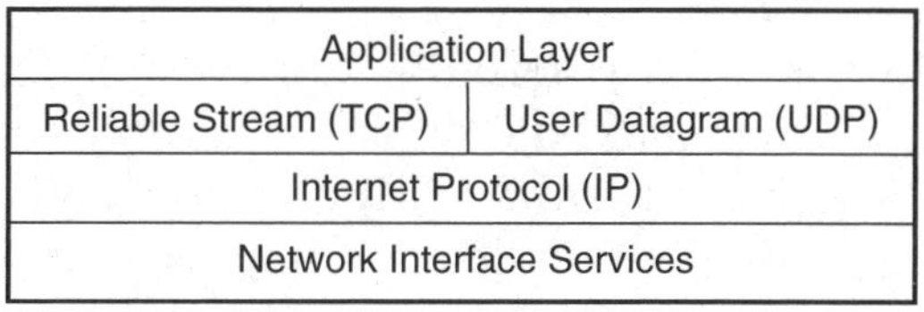

Application Layer	
Reliable Stream (TCP)	User Datagram (UDP)
Internet Protocol (IP)	
Network Interface Services	

Nodes in different networks can have common names. Think of each network as a state and each node as a city. Two cities can have the same name as long as they are in different states. But to find the right city, you need to know which state you should be looking in.

The internet protocol (IP) layer deals with delivery of data packets. It provides packet processing rules, identifies conditions for discarding packets, and controls error detection and error message generation.

Data delivery, concurrency, and sequencing are handled by the Transmission Control Protocol (TCP) layer. TCP also handles connection to applications on other systems, error checking, and retransmission. TCP uses connections between two points, not individual endpoints, as its fundamental concept.

TCP/IP uses a three-way handshake to ensure synchronization between two endpoints. The requester sends a synchronized signal and an initial sequence number to the destination endpoint. The receiver receives the synchronized signal and sends back acknowledgment, sequence number, and synchronization signals. Upon receiving these signals, the requester sends the acknowledgment back to the receiver. This handshake is necessary to ensure that messages are not lost, duplicated, or delayed.

How did this "dated" technology become so popular? As companies began to demand open networking protocols, they turned to IP, and it became a de facto standard. In addition, computer processing power fueled the growth. Switches, routers, and hubs became cheaper. Because of advances in router technology and the use of domain name services (DNS), the task of administering IP addresses has become largely automated. However, the protocol's limitations become all too apparent when bandwidth isn't available.

5.2.1 TCP/IP Addresses

Each station, whether client or server, must have a unique IP address. Client workstations have either a permanently assigned address or, for dialup connections, an address that is assigned dynamically for each session.

Under the current version of IP (IPv4), an Internet Protocol address is 32 bits long, divided into a network number and a host number. IP addresses are written as four sets of numbers separated by periods—for example, 140.170.42.65.

DNS addresses are user-friendly—for example, DLTDewire@aol.com. IP addresses are mapped to English names for nodes and vice versa.

The 32 bits are divided differently depending on the class of the address, which is based on the number of hosts that can be attached to the network. Thus, the more bits used for the host address, the fewer remain for the network address.

For example, a class A address uses 7 bits (the first of which is always zero) for the network number and 24 bits for the host number. A class B address sets the 2 high-order bits of the network portion to 10 and uses 14 bits for the network portion and 16 bits for the host number. A class C address, with high bits of 110 in its network portion, uses 24 bits for its network number and 8 bits for each host number.

IP network addresses are assigned by a central authority, such as the Internet Network Information Center (InterNIC). Administration of node numbers is done locally.

TCP/IP also requires configuration of the host computer. A TCP/IP host machine needs to know its own host name, the name of its network domain, the subnet mask for its network, the IP addresses of any gateways (necessary for communicating with other networks), and the IP addresses to DNS servers to resolve names to IP addressees.

Configuring TCP/IP means going to each local machine and entering this information into local files on each system. This is not too burdensome if the network consists of only a few dozen nodes, but it is overwhelming if the network consists of thousands of nodes.

To help manage and assign IP addresses, two protocols were developed in the mid-1980s: Reverse Address Resolution Protocol (RARP) and the Bootstrap Protocol (BOOTP). The Internet Engineering Task Force (IETF) established the Dynamic Host Configuration (DHC) working group in 1989 to develop techniques for configuring hosts dynamically. The resulting proposal, Dynamic Host Configuration Protocol (DHCP), builds on the existing BOOTP protocol and supports all BOOTP vendor extensions. These three assignment methodologies are discussed in Sec. 10.1, "IP Address Management."

5.2.2 IPv6

As the Internet explodes in popularity, the IP portion of TCP/IP, which deals with addressing, is approaching its limits. The 32-bit address size of IPv4 simply cannot provide enough name spaces to handle probable demand past the year 2003. In addition, the routing tables are becoming too bulky to be of quick use.

To prepare for this demand, in late 1991 the Internet Society's Internet Architecture Board began crafting the next generation of the Internet Protocol, with as its goal the connection of 1 quadrillion (10^{15}) computer networks. Ultimately, in 1994, three proposals hit the IETF, which blended the proposals together into IPng (IP next generation) and then renamed it IPv6. (I don't know what happened to Version 5!)

To minimize the impact of the Internet explosion while work was being done on IPv6, the IETF developed a new technology to reduce the size of routing tables, called Classless Interdomain Routing (CIDR, pronounced "cider"). It reduces routing table growth by having routers advertise groups of routes to networks rather than every individual route. This aggregation can be carried to higher and higher levels: Once routers advertise the collections of routes they "know," the next level up in the routing hierarchy can also advertise aggregations of routes. Internet service providers could aggregate their subscriber networks behind a single routing table entry, shrinking the routing table even further.

The specification for IPv6 was released in early 1997, and vendors began rolling out products that support IPv6 by mid-1997. IPv6 includes a variety of enhancements, including simplified header format, quality-of-service capabilities, and expanded addressing routing. Its autoconfiguration capability allows new IP address nodes to be configured with minimal intervention from network administrators. Security measures are mandated in IPv6. Both authentication and privacy are implemented as extensions to the main IP header.

IPv6 supports an expanded number of network addresses because its address space is expanded to 128 bits. It uses a format of eight 16-bit integers separated by colons. Each integer is represented by four hexadecimal digits. Although the IPv6 header is larger than that of IPv4, it is fixed-length, has fewer fields, and should make routing more efficient, since routers will have less processing to do per header. A number of options are placed in optional headers, most of which are not examined or processed by any router in the packet's path.

IPv6 enables packets belonging to a particular traffic flow to be labeled with a sender's request for special handling. The sender assigns the same flow label to all packets that are part of the same flow. A flow may comprise a single TCP connection or multiple ones. This ability helps support real-time delivery of video and audio traffic.

IPv6 implements an IPSec (IP security protocol) security layer which secures anything using User Datagram Protocol or TCP. IPSec, designed by the IETF, simplifies security management and exists below the transport layer, so it is transparent to applications and users. IPSec supports the perimeter-security model, which applies security measures to everything crossing the perimeter of the network. Traffic within a company or workgroup moves freely without incurring any security overhead.

IPSec uses two headers. The Authentication Header provides source authentication and integrity to the IP datagram. The Encapsulating Security Payload (ESP) header is used to establish confidentiality. Authenticating the IP datagram's source protects against attacks, but ESP prevents an attacker from looking at sensitive information.

Upgrading from IPv4 to IPv6 is a multiple-step process, but IPv6 is not backward-compatible. IP stacks will need to be upgraded for client and host machines. DNS servers and routers will also need to be upgraded and DNS servers refitted with software that can handle the larger addresses. Routing in IPv6 is almost identical to IPv4 routing with CIDR except for the effect of the larger address size. Upgrading the router software is simple, but more RAM may be needed in the routers because the routing tables are larger. Some IPv6 addresses can be generated by appending 96 zero bits to an IPv4 address. This will facilitate tunneling IPv6 packets through an IPv4 network, since the zeros can be added to or removed from the IPv4 address.

If an organization converts from IPv4 to IPv6 in phases, dual stacks will be required on existing hosts to tunnel IPv6 packets over to IPv4, as shown in Fig. 5-2. Tunneling will allow both protocols to exist. IPv6 packets are tunneled by encapsulating the packets in IPv4 datagrams and routing them over the IPv4 network.

Many organizations may elect not to upgrade, since IPv4-based products already provide many of the benefits IPv6 offers. The main reason to upgrade is addressing, and there needs to be a clear business case for utilizing these larger addresses before upgrading makes sense. In addition, there is also growing concern that IPv6 will not be the answer to the addressing problem. IETF is already working on IPv8, and there is even talk of an IPv10.

5.3   HTTP

One of the other important standards used by the Internet is the communication protocol used by the clients and servers—Hypertext Transfer

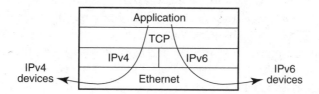

Figure 5-2
Dual-stack machine

Protocol (HTTP). HTTP is used for moving documents around the Internet. HTTP was designed to move long sessions with plenty of data over a single session. In short, HTTP was not designed for its evolved use or level of usage.

Consequently, what happens is that HTTP 1.0 consumes resources on servers, clogs routers with excessive traffic for the amount of data actually transmitted, and produces long waits for users. As a result, organizations are buying server after server to handle the excess sessions to retrieve pages and bandwidth and router capacity to handle spiky traffic patterns.

HTTP 1.1, a proposed standard RFC (2065) by the World Wide Web Consortium (W3C), has two major improvements to address those shortcomings: persistent connections and command pipelining.

A persistent connection uses a request/response sequence that looks identical to that of HTTP 1.0, but the communication happens on the same TCP connection rather than Default 4, which serializes requests. The new approach eliminates the TCP connection starts and stops on both the client and the server. That results in improved packet performance and higher performance on the server, which does not have to manage the session overhead.

Since the protocol is using only one connection, the designers of HTTP 1.1 also changed the clients and servers to pipeline the communications. Instead of waiting for a response to arrive before issuing a new request, the protocol issues several requests at once. The responses are still serialized.

Early tests on HTTP 1.1 had a total data reduction of one-third. Once major browsers support HTTP 1.1 and key servers are converted, products that implement this newer version should begin to appear on the market. This evolution should be sparked by organizations faced with buying more routers, servers, and bandwidth to service their increased usage, both by internal users and by external visitors.

A W3C working group is already looking to the HTTP-NG (Next Generation) project. Members of the group feel that the HTTP protocol has evolved so much, in both size and complexity, that a complete redesign is in order.

HTTP-NG would propose a new architecture based on a distributed object-oriented model. The W3C working group feels that this approach would provide the necessary flexibility to handle future Web needs without being encumbered by the backward-compatibility issue.

For more information on HTTP 1.1, see www.w3c.org/pub/WWW/Protocols/HTTP.

5.4 Internet Administrative Bodies

One of the very interesting things about the Internet is that it is owned by no one. However, it still requires administration. One of the most visible administration areas is domain names—the alphabetic part of Internet addresses.

Internet Network Information Center (InterNIC) is an administrative organization that is currently responsible for allocating domain names and IP addresses. InterNIC registered its 1 millionth domain name in mid-1997. InterNIC (www.internic.net) was originally funded by the National Science Foundation and is now run by Network Solutions Inc. (NSI), which currently holds the monopoly on second-level domain names, such as "aol" in "aol.com." However, NSI's contract expires in March 1998.

Funding will also soon run out for the Internet Assigned Numbers Authority (www.isi.edu), based at the University of Southern California, which handles top-level domains such as .com and .edu.

The Commerce Department invited public input on the future of domain name registration and administration. [Those comments are posted on the National Telecommunications and Information Administration (NTIA) Web site, www.ntia.doc.gov.] And comments it got. The stakes here are high. More than 3000 new domain names are added every day, and most of them end in .com. Just five years ago, most of the 7500 registered domain names ended with .edu or .gov. This is becoming an issue of control similar to phone number control.

The controversy is centered around several related questions:

■ *Who should be responsible for assigning and registering domain names?* Should it be InterNIC exclusively? Should there be new registrars? Should an international governing board be created—and if so, how would the membership of that board be determined?

■ *Should the roster of top-level domains be expanded to include additional types?* As the Internet is growing, do we need to further categorize sites?

■ *What is the role of government?* Which government should it be, considering the Internet's international presence?

■ *What are the legal implications if one registrar holds a monopoly?* These might include trademark infringements and antitrust violations.

■ *Should there be a global trademark policy?* If so, who would create an international process to resolve disputes over domain names—and who would oversee it?

■ *How should country code suffixes be handled?* Because of the Internet's U.S. roots and the early U.S. dominance, "us" is not used as a country code suffix. Who decides if it should be or not?

5.4.1   Administrative Proposals

Network Solutions Inc. (www.netsol.com) has proposed that it control .com and allow new registrars to administer other top-level domain names. This idea has been greeted with possible Justice Department antitrust action for being a monopoly.

A nonprofit group called the American Registry for Internet Numbers (ARIN) has been proposed that would administer and register Internet Protocol addresses (IP addresses) to locations currently managed by NSI. (See its Web site for more information—www.arin.net.)

The Interim Policy Oversight Committee (IPOC) is recommending an unlimited number of domain-name registries and new top-level domain names—.firm, .store, .web, .info, .nom, .art, and .rec. The IPOC (www.gtld-mou.org) also calls for a governing worldwide registration and setting up a nonprofit association in Switzerland. It also proposes that the World Intellectual Property Association moderate trademark disputes.

But IPOC is not without its critics, who say that its policies are bureaucratic and controlling. Unlimited numbers of registries would be very confusing and could lead to trade-name infringements by companies that register multiple domain names (for example, .com, .store, .firm, and .nom).

Smaller groups of concerned individuals are also making their proposals. Iperdome suggests changing the current top-level domain names to .com.us, .net.us, .org.us, and so on, and also suggests adding a personal-domain address (.per), which would eliminate the need to change domain names and addresses when an individual changes Internet service providers. Enhanced Domain Name System calls for free, open-market competition and domain-name registration.

It ultimately comes down to government versus private sector. The Internet is a network of private networks, and as such, how much governmental control should there be? If the private sector can manage the network itself, it should be able to create a self-regulating environment. It is possible that domain names will evolve into directories—problem solved.

One thing is guaranteed: There will be changes, and they will happen soon.

5.5 How to Access the Internet

To access the Internet, a user's computer needs a modem, software, and access to the Internet through an access provider, which can be a major on-line service such as CompuServe or America Online; an Internet service provider (ISP) such as UUNET, NetCom, or one of the telephone companies; or any of hundreds of smaller ISPs throughout the country. Access providers generally offer unlimited access for a fixed rate per month and vary in what add-ons they provide.

The modem should at least transmit at 28,800 bps (bits per second). Modems with lower transmission speeds will work, but they are noticeably slower.

Software may be supplied by the service. The on-line services usually provide an E-mail program, a Web browser, and possibly an Internet Relay Chat (IRC) program. The E-mail feature can be used to send a message to anybody in the world who subscribes to an Internet or on-line service. A Web browser lets a user log onto the World Wide Web and view all the Web pages that are available. The Internet Relay Chat lets a user conference interactively with people all over the world via hundreds of IRC channels on every subject conceivable.

ISPs may expect the user to get his or her own browser, such as Microsoft Internet Explorer or Netscape's Navigator or Communicator. Both have E-mail capabilities. ISPs may also offer to host Web pages for subscribers.

If a user only browses the Web, there is little lost in starting with one service and switching to another. However, switching an E-mail address is not so easy. Most ISPs do not forward mail once an account is closed. A user who expects to use or publish an E-mail address and would like it not to change when he or she changes access providers should register a domain name with InterNIC. In addition, the user will have to use an ISP that supports unique domain names, but most do, and some will even handle the InterNIC registration. Once a domain name is registered, if a user changes access providers, the domain name gets transferred to the new ISP.

CHAPTER 6

World Wide Web

What started the explosion of Internet use was the development of user-friendly graphical interfaces. The first of these was Mosaic, which was created by the University of Illinois National Center for Supercomputing Applications and released on the Internet in early 1993. Mosaic allowed users to navigate the Internet with a point-and-click interface. It also made use of hypertext (links to related information) so that the users didn't have to know the file address of the information they wanted to "browse." All a user did was click on "hot" text, and the software took over by pulling up the screen of information. Also, the software remembered where the user had been, so that the user could get back to previous screens of information within the current session. Users could follow their own thought processes ("I would like to know more about that" and "What does that mean?") rather than a menu of predetermined choices.

The next evolution in Internet software resulted in the World Wide Web (WWW or the Web). The Web is an Internet service that links documents locally and remotely. The Web is the hottest technology to come out of the Internet. Everyone talks about Web pages, and www.something.com

or something.com appears in every ad in print or on TV. Today the terms *Web* and *Internet* are used synonymously.

A Web document is called a Web page, and links in the page let users jump from page to page (hypertext), whether the pages are stored on the same server or are located on servers around the world. The pages are accessed and read via a Web browser such as Netscape Navigator or Microsoft Internet Explorer.

The Web has quickly become the focus of Internet activity because a Web browser provides such easy access to Web pages, which can contain both text and graphics. This point-and-click interface provides a window into the largest collection of on-line information in the world, and the amount of information is increasing at a staggering rate. The Web is also turning into a multimedia delivery system as new browser features and browser plug-in extensions, coming at a dizzying pace, allow for audio, video, telephony, 3-D animations, and videoconferencing over the Net.

Just about the time users grew accustomed to accessing information (text, audio, video, etc.), newer browsers were released that support the Java language and ActiveX controls, which allow applications of all varieties to be downloaded from the Net and run locally.

6.1   Evolution of the Web

The World Wide Web was developed at the European Center for Nuclear Research (CERN) in Geneva from a proposal by Tim Berners-Lee in 1989. It was created to share research information on nuclear physics. In 1991, the first command-line browser was introduced. By the beginning of 1993, there were 50 Web servers, and the Voila X Window browser provided the first graphical capability for the Web. In that same year, CERN introduced its Macintosh browser, and the National Center for Supercomputing Applications (NCSA) in Chicago introduced the X Window version of Mosaic. Mosaic was developed by Marc Andreessen, who later became a principal at Netscape.

By 1994, there were approximately 500 Web sites, and by the start of 1996, there were nearly 15,000 in the United States alone. Today, there are hundreds of thousands of Web sites, with new ones coming on-line at a staggering rate of four per hour. In 1994, there were 5000 home pages. In 1997, there were 5000 created daily. Nearly 8 million people access the Web each day. There are 66 million public pages on the Web, and 1.1 billion are expected by 2000.

The initial applications built using Web technology were marketing-oriented. Web sites (accessed by home pages) provided customers—existing and potential—with information about products and services. The Web became a way to communicate with potential customers without having to reach out and find them—they could find you on the Internet.

6.2   The Web and HTTP

Hypertext Transfer Protocol (HTTP) is used to connect to servers on the World Wide Web. Its primary function is to establish a connection with a server and transmit Hypertext Markup Language (HTML) pages to the client browser. Addresses of Web sites begin with an http:// prefix. Most Web browsers default to the HTTP protocol, so it is usually not necessary to type it in.

HTTP version 1.0, which is currently used, adds considerable overhead to a Web download. Each time a graphic on the same page or another page on the same site is requested, a new protocol connection is established between the browser and the server. In HTTP version 1.1, a persistent connection allows multiple downloads with less overhead. Version 1.1 also improves caching and makes it easier to create virtual hosts (multiple Web sites on the same server). In order to obtain these advantages, both browser and server must be upgraded to the new version.

It is expected that the HTTP protocol will undergo several revisions in an attempt to manage the escalating traffic on the Internet. However, many believe that only entirely new protocols will be able to meet the demand.

6.3   The Web and HTML

The fundamental Web format is a text document embedded with HTML tags that provide the formatting of the page as well as the hypertext links [uniform resource locators (URLs)] to other pages. HTML codes are common alphanumeric characters that can be typed with any text editor or word processor. Numerous Web publishing programs provide a graphical interface for Web page creation and automatically generate the codes. Many word processors and publishing programs also export their existing documents to the HTML format, and thus users can create Web pages without learning any coding system. The ease of page creation has helped fuel the Web's growth.

HTML pages are static. Dynamic HTML allows a Web page's format or layout to be changed without reloading the page—the page becomes active. See Sec. 8.2.3 for more information on Dynamic HTML.

Figure 6-1
How browsers
retrieve Web pages

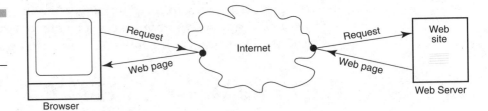

Figure 6-1
How browsers
retrieve Web pages

6.4   Web Browsers

World Wide Web client programs called Web browsers allow users to browse the Web. To begin, a user types in the URL (address) of a Web site. The home page of that site is downloaded to the user's computer, as illustrated in Fig. 6-1. The user clicks on the hyperlinks created on that home page (and subsequent Web pages) to access the information available via that Web site, either locally or located on another Web site.

A home page is created for each site, with links to other documents locally and throughout the Internet.

Web documents are structured with format codes and hypertext links using HTML. These links can include links to text data or multimedia data. For example, by clicking on areas of the screen, a user can see (or hear) a product and in some cases make the product rotate to see it from all angles, read about its features, download sales literature, and leave an address to have a local dealer call.

Browsers also have a bookmark feature (also called Favorites) that allows a user to store references to favorite sites. To revisit a bookmarked site, the user selects it from the bookmark list instead of typing the address.

An Internet "year" (or a Web "year," for that matter) is about 4 to 6 months because new capabilities that can be programmed into Web pages are released, followed by new releases of Web browsers to support them. Since there is considerable competition to have the "coolest" Web site, Web pages are reprogrammed with the new functions as fast as they are released.

Web browsers are discussed in more detail in Sec. 9.2.

6.5   Web Sites

Web pages are maintained at Web sites. When a user accesses a Web site, generally the first link is to the site's home page, which is an HTML document that serves as an index, or springboard, to the site's contents.

Large organizations create and manage their own Web sites. Smaller organizations often have their sites hosted on servers run by their Internet service providers.

6.6   Web Hosting

Some Internet service providers and Internet access providers offer to host Web sites on their computers as part of their services. For this Web hosting service, the outside companies are typically charged on the basis of equipment and transmission capacity used.

A Web service provider provides Web pages on its computers. It may also provide additional services, such as design help and usage statistics, or it may just provide the hard disk storage and allow the outside company to do the rest. A Web service provider is not necessarily an Internet service provider.

6.7   Web Servers

A Web server is a server containing Web pages and other files which is on-line to the Internet 24 hours a day. A Web server can host multiple Web sites. It supports the HTTP protocol. Web servers are discussed in more detail in Sec. 9.3.

A secure Web server is a server on the Web that supports one or more of the major security protocols, such as Secure Socket Layer (SSL), Secure Hypertext Transfer Protocol (S-HTTP), or Private Communications Technology (PCT). These security protocols are discussed in Sec. 11.4.

6.8   In-House Webs—Intranets

The World Wide Web has spawned the intranet, which is an in-house, private Web site for users within the organization that is built using Internet technologies. The intranet is used for internal document sharing, data access, and transactions. An intranet is protected from the Internet by a firewall that lets intranet users out to the Internet, but prevents Internet users from coming in.

If an organization can piggyback on its existing TCP/IP network, there is no need for a dedicated network.

6.9   Web-based File Systems

Sharing large files is a common task on intranets, and the Internet is also increasingly being used for that purpose. To address this problem, SunSoft developed a Web version of its Network File System (NFS) distributed file system. Microsoft followed suit with the Common Internet File System (CIFS).

Support for WebNFS and CIFS is beginning to appear in Web browsers—or at least be promised for the next release. Currently the only server that supports WebNFS is Sun's Netra NFS Internet server, and only Windows NT servers include support for CIFS.

6.9.1   WebNFS

Sun created NFS to solve file-sharing problems for users and network managers. Before NFS, users had to download entire files from the server, a time-consuming process that also wasted local storage. Or, if the server supported their clients, users could log on and read from and write to files, but this forced them to remain connected throughout the job.

In contrast, NFS clients can "mount" a server file system without remaining connected to the server. To the user, the file system appears to be on a local disk. NFS moves file data automatically as needed. The high speed of Ethernet, which provides performance that sometimes exceeds that of the local disk, makes this process transparent to the end user.

WebNFS is a Web-based outgrowth of NFS. Among other things, it adds firewall compatibility and a global naming scheme. In addition, it provides a safer, more efficient way to download files than current Internet protocols offer.

WebNFS-enabled servers and browsers allow access to Web pages as much as 10 times faster than the standard HTTP protocol. Unlike HTTP, which drops the connection after each tiny file is downloaded, only to reconnect for the next one, WebNFS downloads multiple files with a single connection. It also provides fault tolerance for large downloads that lose their connection in midstream.

WebNFS was designed as a replacement for HTTP, the server-based transfer protocol, and ftp (File Transfer Protocol), the client-based protocol. Unlike HTTP or ftp, WebNFS allows a user to access a specific page of a file instead of downloading the entire file. If the connection is broken during a file download, WebNFS also keeps track of where the transmission ceased and starts at that point after the connection is reestablished.

Ftp downloads any kind of file, from long text documents to tiny graphics on a Web page, separately through a single TCP connection. Once the file is downloaded, the connection is dropped and must be reestablished for the next file. A WebNFS client can download all files with a single TCP connection. This should shorten the time it takes a Web page to appear and allow a server to handle more clients simultaneously.

6.9.2   CIFS

SunSoft's competitor to WebNFS is CIFS. This new dialect of Microsoft's Server Message Block (SMB) protocol performs many of the same functions as WebNFS. CIFS provides LAN Manager—style file sharing that uses SMB and NetBIOS over TCP. SMB is an intranet protocol for sharing files, printers, serial ports, and communications abstractions (such as named pipes and mail slots) between computers. The file-sharing protocol built into Windows NT and Windows 95 is based on SMB.

Each SMB server makes a set of resources available to clients on the network. A shared resource could be a directory tree, named pipe, or printer, for example. The client considers the server to be the sole provider of the file (or other resource) being accessed. The SMB protocol requires server authentication of users before file accesses are allowed, and each server authenticates its own users. A client system must send authentication information to the server before the server will allow access to its resources.

6.10   Web Cookies

Did you ever wonder how a Web site knows which links you've already looked at? Or opens your shopping cart with the items you bought the last time you went shopping on-line? Or knows what your preferences are? It's easy—the Web site sent you a cookie, and it's sitting in your machine.

HTTP cookies—usually referred to just as cookies—are text files of information sent back and forth between a Web site and the Web browser. What's in the cookie is a string of characters unique to that user that is generated by a visited Web site. Later, when the user returns to that same Web site, it grabs the cookie from the hard drive and remembers who the user is.

Cookies allow HTTP to remember things. Remember that HTTP is a stateless protocol: A browser makes a connection and sends a request, and the server responds to the request by sending a file. Then the connection is broken. The server has no way of remembering the transaction. Cookies take care of that. Cookies are possible because an HTTP transaction involves a request and a response. Cookies are exchanged between the server and the browser through the HTTP headers.

When a user logs onto a Web server, the server interrogates the user's machine by looking at its cookies. The information contained in the cookies is accessed by a server every time the user accesses that server. A cookie is sent to the Web browser along with the rest of the HTML document that is requested. On the next visit to that Web site, the cookie is transmitted back to the server, which may then send an updated cookie.

To learn more about cookies, see www.cookiecentral.com, Malcolm's Guide to Persistent Cookies Resources (www.emf.net/`mal/cookiesinfo.html), or Netscape's Guide to Cookies (home.netscape.com/newsref/stdcookie_spec). A site managed by the Center for Democracy and Technology (www.cdt.org) demonstrates the kinds of data that are revealed when a user logs onto a Web site.

6.10.1   Worry about Cookies?

Cookies are safe. They can be no larger than 4K bytes and can be read only by the Web site that sent them. A cookie can't read anything on the hard drive it's written to. However, it is an invasion of privacy. Once a Web site has sent a cookie to a machine, it can track user movements through the site. It will know that the user has been there before. It will know which ads the user looked at and either not show those again or show similar ones. A cookie can actually replace a site log-in, which forces users to enter a user name and a password every time they log onto a site. The data is stored in a cookie instead.

If invasion of privacy is a big concern, consider what a Webmaster already knows about a visitor. Every time a Web page is downloaded, the browser sends the Web server information about what Web browser and operating system is being used, the URL of the last page visited, and the IP address of the computer being used. All that on the click of a link.

What should be of concern is whether these cookies are encrypted or not. If the cookie data can be edited, then the server could be given the wrong data—and what if it is someone else's credit card?

For now, cookies seem innocuous. As the Web becomes more commercial, that feeling is bound to change.

6.10.2 Cookie Control

Most Web servers create cookies that automatically self-destruct if they are not used within a set period of time. Navigator can store only up to 300 different cookies on a hard drive, with up to 20 cookies per domain name or server.

Users can stop the flow of cookies. The user can have the browser warn the user each time a site tries to send a cookie (in Netscape Navigator, Options: Network Preferences: Protocols: Show an Alert Before Accepting a Cookie, and in Internet Explorer, View: Options: Advanced: Warn before accepting cookies). Navigator's cookie file, MagicCookie, can be deleted. To find cookies on a hard drive, use Explorer to search for the word *cookie*. Navigator places all cookies in one file; Internet Explorer puts each cookie in a separate file with the domain name as the filename in a folder. These files can be accessed with a text editor or WordPad to see which Web server (by domain name) sent the cookie.

To manage cookies Altus Software Marketing (www.wizvax/kevinmca) offers IEClean for Internet Explorer and NSClean for Navigator. They won't stop sites from delivering cookies, but they do allow a user to delete unwanted items once the browser is closed. Pretty Good Privacy (www.pgp.com) offers PGPcookie.cutter, which is a browser plug-in that doesn't block unwanted cookies but does prevent servers from looking at the information. CookieMaster was available free from ZDNet and quickly became one of its most popular downloads.

Proposal RFC 2109 from the IETF will eliminate anonymous cookies and will allow users to accept the cookies they want and reject those they do not want. Internet watchdogs such as Electronic Frontier Foundation (www.eff.org) are also concerned about the invasion of privacy that cookies make possible.

6.11 Webcasting

Webcasting mixes the automatic notification of E-mail with the efficiency of broadcasting. It confers the benefits of the Web by providing fast information access but shields users from Web limitations such as huge searches. News, entertainment, and other kinds of companies create channels with Webcasting programs that allow them to deliver their content directly to the desktop. So in a Webcasting program, there is a channel created by *The New York Times*, another by CNN, and yet another by

Reuters, to name a few. Users subscribe (for free) to get information from each of those channels.

Webcasting is also called push technology, which is discussed in Part 4.

6.12 Standards-Setting Bodies

The key to the growth of the Internet continues to be adherence to standards. There are currently three standards-setting bodies that are focusing exclusively on the Internet and the World Wide Web. The Internet Engineering Task Force (IETF) also has subgroups that focus on the Internet.

6.12.1 World Wide Web Consortium

The World Wide Web Consortium (W3C), based in Cambridge, Massachusetts, is a consortium run jointly by the MIT Laboratory for Computer Science in the United States, the National Institute for Research in Computer Science and Control in France, and Keio University in Japan. W3C is trying to drive standards into the Internet through the Webcore (World Wide Web Core Development) project. W3C comprises more than 150 companies. The membership includes many of the industry's leading companies, such as AT&T, Adobe Systems, Digital Equipment, Sun Microsystems, Novell, HP, Microsoft, Netscape, and Spyglass.

W3C was created to develop common protocols that enhance the interoperability and promote the evolution of the World Wide Web. Services provided by the consortium include a repository of information about the World Wide Web for developers and users, reference code implementations to embody and promote standards, and various prototype and sample applications to demonstrate use of new technology.

W3C defines technical specifications for intranets and the public Internet. The following standards are currently being reviewed by W3C:

- Dynamic HTML allows Web pages to change without reloading the page.
- Extensible Markup Language (XML) is a subset of the Standard Generalized Markup Language for creating custom HTML tags.
- Channel Definition Format (CDF) is an XML extension proposed by Microsoft to define content and formats for pushed or Webcast content.

The membership has also spearheaded the Joint Electronic Payments Initiative (JEPI) for electronic wallets, which are virtual wallets for holding such things as electronic cash and credit card numbers. JEPI provides a standard for different types of payment methods and protocols.

For more information, visit W3C's site at www.w3c.org.

In some ways, the objectives of W3C are not very different from those of the IETF, except that its members have commercial interests—the IETF's members are volunteers and tend to come from academia. W3C focuses on the narrower Web, whereas the IETF focuses on the broader Internet.

6.12.2   Desktop Management Task Force

The Desktop Management Task Force (DMTF) is an ad hoc network-management standards-setting group. DMTF (www.dmtf.com) was initiated by Intel in 1992 and developed the Desktop Management Interface (DMI), which is a management system for PCs. DMI provides a bidirectional path to interrogate all the hardware and software components within a PC. When PCs are DMI-enabled, their hardware and software configurations can be monitored from a central station in the network.

DMI is a complete management system, but it can coexist with Simple Network Management Protocol (SNMP) and other management protocols. For example, when an SNMP query arrives, DMI can fill out the SNMP management information base with data from its management information files. A single workstation or server can serve as a proxy agent that would contain the SNMP module and service an entire LAN segment of DMI machines.

The DMTF Management Standards Evolution roadmap released in mid-1997 describes how the DMI standard will work with the new object-oriented Common Information Model (CIM) specification—also known as the Hypermedia Management Schema—from the Web-based Enterprise Management group and Web-based technologies. Updates to the specification were released in late 1997, and products based on these standards should be available sometime in 1998.

CIM supports popular object and Web-based technologies such as the Object Management Group's Common Object Request Broker Architecture (CORBA), Microsoft's Component Object Model (COM), and Sun Microsystems' Java Management Application Interface (JMAPI). It also is the data model for the Microsoft-backed Web-based Enterprise Management initiative.

CIM will let administrators gather data from DMI-enabled, SNMP-based, and CMIP-compliant devices and systems to provide a complete picture of what is happening in a networked systems management environment. See Sec. 14.3.3 for more information about CIM.

DMTF is also working on organizing cost-of-ownership (COO) standards groups to assist organizations in collecting and tracking information on their IT assets. The COO groups will work on all aspects of PC costs, including software, hardware, and support. Some of the hardware studies have been completed, and the group is now focusing on software.

In January 1998, the COO group released its first iteration of a system-based cost-of-ownership solution which is based on a local database that collects, stores, and transmits business and asset management information about the desktop. See www.dmtf.org/work/cost for more information.

6.12.3   Web-based Enterprise Management

Seventy vendors, including Intel, Cisco Systems, Compaq, and Microsoft, have formed the Web-Based Enterprise Management (WBEM) group to integrate Web perspectives into management standards. They are developing standards that will overhaul all aspects of systems and network management.

Their proposal, also called WBEM, contains three components. The Common Information Model represents management information across all components: hardware, operating system, applications, and databases. This was presented to the Desktop Management Task Force for approval. The second is a protocol called Hypermedia Management Protocol that accesses the information, which must be approved by the Internet Engineering Task Force. The third, called Hypermedia Object Manager, is a management solution that presents the information in a Web browser. See Sec. 14.3.3 for more information about WBEM.

These specifications allow management hardware and software to share data. They can be viewed at www.wbem.freerange.com.

6.13   Where Is the Web Headed?

Many believe that the Web signifies the beginning of the real information age and envision it as the business model for the twenty-first century. Others consider it the "World Wide Wait" right now, as surfing the Net via modem using an Internet service provider (who is always scrambling to keep up with demand) is often an exercise in patience.

It seems that everyone has some vested interest in the Web. The telephone and cable companies want to provide users with high-speed access to the Internet from home. Existing Internet service providers want to gain market share. IT managers are concerned with vulnerability when the internal network is connected to the Internet. The publishing industry is perplexed over how to manage its copyrighted material on a medium that can send a document all over the world in a split second. Software vendors are scrambling to make their products Web-compatible. Web-based search engines are trying to differentiate themselves. Netscape and Microsoft are trying to add features to their browser products to make them more attractive to users as well as developers—so they are not considered "just browsers." Hardware vendors are debating whether the network computer (NC) will replace the PC.

It is safe to say that nothing in the computer/communications field has ever come onto the scene with such intensity and speed. Can the Internet bear up under the load? Will the Internet remain unregulated? Will the use of the Web fizzle out? Or is the Web the marketplace and communication mode for the twenty-first century?

CHAPTER 7

Intranets and Extranets

First, some definitions. From the technologist's perspective, an intranet is simply the use of Internet protocols for an internal communications network. The same communications technology used to create the Internet is also used to link a company's internal networks.

The easiest way to think of an intranet is as an IP network separated from the Net by some device or software—a firewall. To jump on this fast-moving technology, some companies have simply called their existing internal networks intranets, whether or not they are fully based on IP technology. The term has been used to refer to internal Web sites, servers, and hosts that can be accessed through a browser. Some companies refer to such networks as "internal Webs." However, most companies use the term *intranet* to describe any device available on an IP network.

An extranet is an intranet that is extended—it allows access from outside the firewall. Since an extranet has as its foundation an intranet, this book will use the term *intranet* to include both except for those areas that are totally different. Most of the extranet focus in this book is in Part 5, "Electronic Commerce."

Applications can be based on the Internet structure, can be internal, or can allow external access. These applications share many traits, such as use of standardization. The term *i-net* will be used to signify the traits that are common to all these types of applications and their architectures.

7.1   Why Intranets?

An intranet's strength is that it is based on a series of extremely pervasive protocols and standards. The most obvious example is TCP/IP, the networking protocol that forms the backbone of an intranet's communication infrastructure. Other examples include Hypertext Markup Language (HTML) and Hypertext Transfer Protocol (HTTP), which have made the World Wide Web what it is today. HTML provides a standard tag language for text files, including the ability to embed graphic images and hyperlinks referencing other files on remote computers. HTTP is the communication method that exchanges HTML files across the network.

Intranets offer many advantages over traditional LANs, including the availability of collaborative groupware and simplified messaging. Intranet software can be inexpensive; Microsoft Internet Information Server is free for users of Windows NT. Browsers that reside on the users' machines are inexpensive or free as well.

Intranets allow an organization to rethink the way applications are designed and installed. Intranets can be used as a vehicle to redesign the way organizations share information. The focus of the communications networks changes from connecting people to providing "the right data to the right people at the right time." Intranets are attractive because they provide a way for companies to mix internal databases with external data and simulation models to create new products. The technology allows companies to differentiate themselves by externalizing internal systems. For examples, Federal Express's package tracking Web application used to be the firm's internal system.

I-net technology can be used to rewrite the rules on how companies adopt new technologies. Competitive pressures, coupled with the lightning-quick acceptance of intranet technology, have forced companies to abandon traditional return on investment (ROI) models and practices. Rather than carefully evaluating bottom-line benefits, corporations are plunging headfirst into the intranet waters. New applications have sprung up in months or even weeks, and top management approval is often assumed rather than requested.

7.2 Evolution of Intranets

It is clear that the interest in intranets has been accelerated by the rapid acceptance of the Web as the de facto public data network and the relative ease of use offered by browsers. Over the last two years, organizations discovered that they could easily set up a Web server, convert documents to HTML, and allow anyone in the organization to have access.

As is so typical of the introduction of new technology into an organization, many intranet projects spring up as grassroots efforts (it looks like fun, it's new, etc.) or as quick and dirty solutions to problems—always with good intentions of stepping back after the fact and assessing (and documenting, of course) what was built. It is beginning to look very similar to the days when PCs were office equipment and were hidden in a closet.

The grassroots scenario goes something like this: An employee was intrigued by the technology. A computer was chosen as a Web server based on availability. Web server software was selected based on what was free or already installed. Security wasn't even considered. Content management was performed by pointing individual authors at particular directories: "Put your HTML files in here, but please don't override or delete anybody else's stuff." Common gateway interface (CGI) programs and integration with the database and legacy systems were coded on an "as-needed" basis using tools familiar to the person doing the coding. Users focused on their own particular area of the organization rather than developing a companywide intranet. More than likely, no one outside the department had access to the site.

The first widespread intranet applications were based on publishing information on an internal Web site rather than distributing it on paper. Suddenly everyone was "intranet publishing"—distributing important business information as HTML pages.

At some point, organizations realized that the small intranets needed to be joined into a companywide intranet, that content was actually being duplicated and shouldn't be, that code to access a corporate database had been written multiple times, that there were no standards in place, that the interfaces all looked different, and that the benefits experienced by one department could be replicated by another department—if that department had known about them. Sounds like the PC revolution all over again, doesn't it?

In 1996, the promise of the intranet—to unify disparate systems into one information architecture—sent many companies scrambling to begin purchasing and developing Web applications. However, in 1997, the excitement started to give way to sobering reality as companies wrestled with deploying a full-scale intranet that provides the initial document sharing, as well as groupware services, decision support, global directory services, site management, and client/server business applications.

Usual targets for intranet implementations were company directories, human resources documentation, and corporate policy statements. While this doesn't sound like much, such automation can have a profound impact on a company with tens of thousands of employees if every update to a document doesn't require generating thousands of copies and then distributing them.

Once companies start intranet publishing, they turn their attention to giving access to the legacy data everyone's been asking for. Employees use a browser to access a corporate database—do queries, generate reports, etc.

Most of the original intranets consisted of isolated Web servers used primarily to support document sharing. It's a big jump to turn these into an enterprise platform with data access and improved security. Infrastructures such as the help desk, software distribution and management, and application guidelines don't exist. Management tools to manage large-scale systems are still not ready. Document publishing is relatively straightforward and easy to implement, and has little impact on the existing infrastructure. But discussion databases, decision support, and business applications add complexity and have a greater impact on the infrastructure.

As companies forge ahead, they turn to applications—not just information retrieval, but applications that do something. They start building intranet-based applications. Suddenly security becomes a major area of concern.

Once a company has set up a secure network with secure transaction capabilities, the next step in the evolution is extranets—making business-critical data available to partners and suppliers. Once a company starts reaching outside its own borders, the next step is electronic storefronts—reaching its clients and customers

All this network traffic strains the network infrastructure, and the company has to upgrade. Once a company has a high-capacity internal network, it can deploy bandwidth-intensive applications that weren't realistic before, such as videoconferencing. At this point, the only roadblock is one's imagination.

Many organizations were burned by the promise of savings with client/server architectures that never materialized. There are savings with intranets—on the client side. But the costs don't disappear, they just shift. Organizations end up investing in both bigger, faster servers and, for their networking infrastructure, capacity, security, and network management tools.

The Web is also reinventing the software industry. Client/server architectures are starting to rely more on powerful host servers and less on overloaded desktop PCs. Vendors, systems integrators, and consulting outfits of all sizes are competing for the business of companies moving to the Web.

7.2.1 Second-Generation Intranets

Some companies are already on their second-generation intranets. These intranets are more interactive, offer higher performance, and are more open than before. Corporations are using them to link their at-home users to sensitive corporate data using tunneling. They tie in distributors and channel partners. Authorized users are allowed to publish directly to their company's internal Web sites from the road—even if the content can't be checked.

Key to these types of intranet uses is that they integrate new and old networking technology by using the latest high-speed switches and private networks and services. By focusing on how private LANs and WANs are alike, companies can blend their conventional networks with the components that are the classic signature of the Internet—a browser, a Web server, and IP transports.

Intranets are a paradox. They can help some users get data faster, but they may be productivity killers for others. An intranet that aids customer service reps in tracking orders for customers is a plus for them—but not for the sales force that might have to use the unfamiliar "tool" while trying to close a deal on the phone. As organizations rethink their intranet design, many are turning to push technology to streamline information delivery. (See Part 4 for more information on push technology.)

Push technology allows an organization to automatically push relevant data to end users as either E-mail or personalized dynamic Web pages. Users don't have to search for data on their own; data basically comes to them. Instead of being faced with hundreds of pages, a user has a starting point. And instead of feeling as if they might have missed something, users feel informed. Instead of pointing and clicking aimlessly for crucial data, they spend their time analyzing their data. In addition, networks aren't clogged as users retrieve Web pages that turn out to be irrelevant.

7.3 Components of an Intranet

As illustrated in Fig. 7-1, an intranet is inside the company's firewall. It uses the same technology the Internet applications use. There is a Web

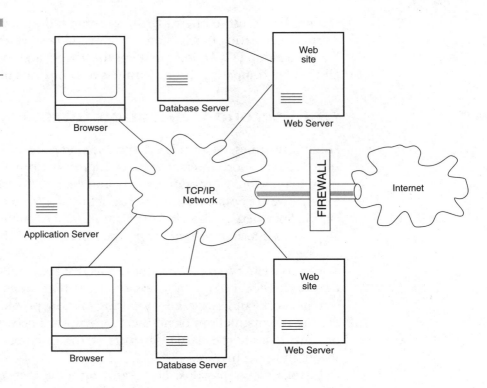

Figure 7-1
Components of an
Intranet

browser on the client machine to access the information or applications within the intranet. The internal network is running TCP/IP. There are Web servers that contain Web sites. Other servers, such as database servers or application servers, will either be linked directly to the Web servers or be accessible to Web servers via the network.

7.4 Reason for the Paradigm Shift

The remarkable things about the Web are that it expanded so quickly (in less than a year!) and that it is a *truly* open system.

Point-and-click hyperlinking allows users to navigate and locate information easily, using standard Web browsers such as Netscape Navigator from Netscape Software, Microsoft Internet Explorer from Microsoft, and

Mosaic variants from National Center for Supercomputing Applications (NCSA), Internet Division of CompuServe, and IBM. The current de facto standards are Netscape Navigator (and Communicator) and Microsoft Internet Explorer.

Programming languages such as Java from JavaSoft can be used to write dynamic applications (applets) that users activate as needed (just-in-time) and that disappear when the application is terminated. Once activated, the applet is downloaded and run directly from the Web browser without the need to maintain a server connection.

Microsoft ActiveX components can also be used to add functionality to Web pages. These are pieces of code that are also downloaded to the client machine but remain there. They will be called up the next time the application is run.

Either way, for users, the browser becomes the single point of interface for access to all internal and external resources.

7.4.1 Open System

Many organizations have been striving to create a truly open system, and most would have predicted it to be impossible, or at best improbable. By the very nature of its architecture, the Web must be truly open, as every site has a different architecture, but all information must be accessible and readable regardless of what client platform asks for it.

Instead of there being different software for different client operating systems and the back-end data sources being accessed, the Web browser does it all. Web browser software sits on the client. When a Web server responds to a client request from browser software, the server doesn't care whether the client is a Windows 95, Windows 3.11, Windows NT, OS/2, or Macintosh client. The Web browser takes care of any and all translations that must be done. It translates requests to the Web server and takes responses and translates them to the client platform.

In addition, since the protocol for the Internet is of necessity TCP/IP (a nonproprietary protocol), there are not multiple network protocols to be concerned with. TCP/IP is in place in most organizations, and it's a mature protocol for networking. HTTP is the client/server protocol used for information sharing on the Internet.

Java, currently the most widely used programming language for Web pages, doesn't lock organizations into proprietary technologies. Java can be considered open (for now) because it is freely and cheaply licensed to any vendor that wants it.

7.4.2   Single Point of Maintenance

Organizations no longer have to configure machines and then distribute new software releases, or updates and patches, to individual machines. Application releases are maintained in only one place—on a Web server. End users access them on an as-needed basis.

In the traditional client/server model (two or more tiers), different versions of client software had to be maintained to support the various operating systems and graphical user interface (GUI) drivers. Any attempts at standardization were impeded by new releases of the products or newer platforms—organizations were trying to standardize on a moving target.

With intranet tools like Java, the server application is written once and the client application is written once. The only thing that needs attention to make the application work across all operating systems is the interpreter—the Web browser.

7.4.3   Manageable Hardware Costs

Client workstations don't need as much horsepower to run only Web-based applications. These very thin clients run only the Web browser and the operating system. The server performs the bulk of data querying and massaging, and of presentation.

Specialized machines, referred to as network computers, sell for about $400, rather than the $1500 for a micro. The devices can connect to a television and access applications and data, via a modem, from a computer network or the Internet. Storage is on the Web server instead of on a hard drive. However, until productivity tools such as word processors and spreadsheets can run as Web applications, most users will continue to need the power of a micro.

7.4.4   Portable

An intranet application is portable, since the popular Web browsers such as Netscape Navigator and Microsoft Internet Explorer are multiplatform. Almost all computer platforms offer some sort of Web browser, which can be used as long as the browsers support the same features. The developer writes the application once and lets users install it on their platform of choice.

7.4.5 Viewed as Minimal Risk

Organizations that have already adopted client/server architectures find that intranets are an extension of that paradigm. TCP/IP is already in place. Adding another front-end tool (the Web browser) is not overwhelming to their users. Learning Java is easy for anyone with a C++ background. User training costs can be lowered if the Web browser is the consistent user front end and query tool.

When a new corporate application is available, the user links to the appropriate Web browser and the new application runs as an executable Java applet, either by downloading and executing a Java applet or as an HTML stream sent from the server or server-based application layer.

Suddenly the idea of "any place, any time" for user access—internal and external users—seems possible and doable. The best part is: It uses inexpensive Web technology.

7.5 Hurdles

For many organizations, internal information dissemination is as big a challenge as providing an external link. Incompatible applications and communications problems often isolate critical information. Interfaces, special reports, and flat-file transfers are some of the ways organizations overcome this gap.

In too many organizations, managers and technicians see themselves as owners of systems and data, rather than custodians. They see the information as an end in itself instead of a service. If this attitude prevails, no matter how successful the intranet implementation is, internal and external customers will not find much useful information.

7.5.1 Network Needs

It is important to assemble the requisite networking gear to support the explosive demand for bandwidth that intranets inevitably produce—a company has to invest in the network.

Corporations with disparate internal networking technologies can combine several mininetworks into a larger intranet. Implementing Web technologies even on this smaller scale would increase employee data access and productivity. Organizations can introduce an intranet incrementally and expand it as needed. Security and management provisions will still need to be expanded as the intranet grows. The complexity and duration of an intranet implementation depend on the number of users that regularly access the network and the flexibility of the technology that is already in place to support additions that provide intranet functionality.

But network investment goes beyond just tying together existing IP networks. And bandwidth capacity becomes a big issue. Enough bandwidth doesn't guarantee success. Planning too far ahead, say by going with asynchronous transfer mode (ATM) for a backbone when there is no ATM in the organization, may actually complicate an intranet implementation. Higher speeds can always be added later.

7.5.2 Centralized Authority

Many corporations are realizing that their intranets have grown into massive, uncontrollable beasts that require new data management strategies. When an organization gets ready to build an enterprisewide intranet, it first needs to find all the existing "autonomous" internal company Web sites. These folks then need to be convinced that they should be brought into a single standard and that the Web group should have centralized responsibility for them. The Web group—a steering committee, a council, etc.—is the governing office for intranets.

One place to start is to publish standards for establishing a company Web site and to use the Webmaster as a central source of input about changes and new developments. In addition, the Webmaster can act as a mediator when intranet deployment is a byproduct of reengineering, which inevitably means conflicts within an organization. And the situation can be exacerbated when the existing IT establishment views intranets as a threat.

One approach is to define and implement Web page authoring policies and guidelines at the onset. While the responsibility for updating information on a Web site falls on the content owner, some companies assign their IT staff to educate employees on Web standards for accessing and publishing pages.

In addition to design guidelines, companies should develop policies for employee use of corporate intranets that cover protection of proprietary information, define inappropriate use, and post regulations. Employees should be careful about posting copyrighted, insensitive, proprietary, and bandwidth-intensive information on corporate intranets. Information that a company deems sensitive should be available only to classified employees, and directions concerning the permitted dissemination of that data should be clear.

Some other rules of thumb are that intranet users should avoid a heavy reliance on multimedia applications in order to minimize bandwidth consumption and should never open an executable file that has come from external sources.

Guidelines about the technological characteristics of all Web pages should also be clearly stated and enforced. IT managers are learning that they must strike a balance between letting users publish freely and imposing standards that keep the network from becoming chaotic. At one end of the spectrum, the procedure might be to have all content changes approved by a content owner in that department. Approved content is sent to a central administrator, who tests the files on a staging server before forwarding them to a remote service provider that hosts the company's content on its servers.

At the other end, users must first submit the content to their department manager. Then IT managers and the Webmaster review it. Users then upload the content to a testing server. Finally, the Webmaster loads the files on a staging server, and custom-written scripts automatically load the content on the live server.

And yet another scenario is to cede control of Web servers, but still oversee content development as much as possible, in an effort to ensure that pages display in all browsers and that multimedia content doesn't suck up too much bandwidth. Each department builds its own intranet site.

7.5.3 Authentication

In an enterprisewide intranet, browser users need to be able to log on to each server or subnetwork. In large corporations, this requires a number of different login names and passwords to authenticate users. Netscape's Lightweight Directory Access Protocol (LDAP) server technology addresses that problem. It provides information about people as well as a storage medium for authentication certificates. Another solution is to use a domain level for the cookie that tracks users, so that a domain cookie can pass the authentication to each server in succession. (See Sec. 6.10 for more information on Web cookies.)

Once the problem of authentication is resolved, a browser can easily become a ubiquitous user interface for a wide range of database applications. One of the more advanced intranet-based data warehousing applications is used by the military.

7.5.4 New Audiences

Thanks to lower-cost Web technologies and a rapidly changing IT mindset, many companies, such as Delta, Levi Strauss & Co., and Tektronix Inc., are starting to roll out intranets that are not just targeting traditional knowledge workers but also reaching places like factories, hospitals, and maintenance yards—workers who have historically not benefited from IT. IT managers have to learn how to design applications and user interfaces for this new audience and determine how much training they need. This broad deployment of intranets to new types of users could bring about big changes in how management values IT investments and how companies communicate and share information more effectively. In today's workplace there are no nonknowledge workers. The people on the front lines are making decisions every day that affect the business, and without access to the information they need, they're going to make bad decisions.

Delta plans to use its intranet infrastructure—called DeltaNet—to automate expensive, paper-based processes such as updating and distributing the massive manuals that mechanics use to repair and maintain aircraft. Currently, these critical manuals are printed and physically shipped to all Delta locations worldwide, an expensive process that can take anywhere from a couple of weeks to six months. Once they're on-line, however, the manuals can be updated once and are available instantly. For the mechanics, DeltaNet means better productivity and the feeling that they are part of the company culture.

San Francisco–based apparel manufacturer Levi Strauss plans to extend its intranet into practically every area of the company. The company's plan, code-named Eureka, uses a collection of kiosks, desktop devices, publishing, and push technology to link workers in 70 countries. Levi's goal is to promote creativity and the sharing of ideas around the world by giving just about everyone in the company access to the intranet.

These companies are the exceptions so far; not every company is pursuing intranet penetration down to the factory floor. IT historically has not looked at all employees as consumers of information. Why the change? It's simple: costs and standards. Companies see in Web technologies a way to roll out systems to a larger audience of users without big capital expenditures or big support bills. In addition, rolling out intranets even around the world becomes easier because IT doesn't have to dictate to local users what kind of desktop hardware to use.

The payoff for rolling out intranets to nontraditional users is likely to be the same as that for Web-enabling knowledge workers: lower administrative costs and smoother, more efficient business processes. Such a rollout also presents unique challenges. Since many of these new users have little experience with desktop systems, for example, how much training will they require, and what type of interface design will work best? Also, in environments where intranet access will be provided via shared kiosks, what's the right ratio of users to machines?

Many companies are hoping to keep training costs for new users low by using a simple approach to application and user-interface design. IT needs to enforce as much consistency and simplicity in the interface design as possible, especially where intranets are being targeted at a wide variety of users. But this doesn't rule out training for new users; it just means different approaches. Some new users will benefit from formal classes, and others from on-the-job instruction from colleagues who have been formally trained—the "train the trainer" approach.

If intranets eventually achieve the predicted broad deployment up and down the corporate hierarchy, one of the major long-term shifts will be in the way line management thinks about IT investments and the way communication happens inside companies. When practically every employee has an intranet-access device, management should stop thinking about IT as an extraordinary capital investment and simply think of it as a cost of doing business. In addition, linking everyone from the top of the organization chart to the bottom in a single communication network may finally help companies achieve flatter, less hierarchical communications and quicker decisions. The technology is perfect to support team-based organizations, a concept that has been discussed since the early 1990s.

7.5.5   Sell the Intranet

But like any new technology, an intranet can go unused unless users have a reason (or enticement) to use it. The idea of information ownership needs to be replaced with one of information sharing. Companies may have to change some of their business practices to promote collaboration among employees. Reward systems tend to foster competition, not teamwork. Intranets prompt teamwork.

IT organizations just can't build the intranet and expect users to come. They must be ready to prompt it, train users, and constantly maintain content. Users need a reason to use the intranet—even if it's just to see what's changed.

A more aggressive way to encourage intranet use is to give employees no alternatives. If data is available only through the intranet, users will have to use the intranet.

7.5.6   User Training

Just as users got comfortable with Windows and PCs by playing games like Solitaire, users are getting comfortable with and skilled on the Web by "wasting" time using a Web browser to read sports or weather, look up stock quotes, or discuss their favorite vacation spot on Usenet.

It's these users who have learned how to compose an effective message in a discussion forum, learned the strengths and weaknesses of the various search engines, and figured out how to phrase a search phrase so as to get a meaningful (and complete) answer.

As organizations roll out large-scale intranet applications, their audience is likely to be users who are less than savvy about electronic communications. Many will be getting Internet access from their desktops for the first time.

Sure, a training class on the organization's browser gets people started, but they need time—and the comfort level that it's OK to spend that time—to get to know what the Internet and the intranet are all about. They need time to experiment.

Training needs to focus on productivity issues: Using E-mail to carry on a discussion to its conclusion. Finding good sources of information, not just a source. How to keep up with updates on intranet sites. How to use push technology.

7.5.7   Intranets and Data Access

Watching the success of Internet browsers, companies are turning to the intranet paradigm to improve and simplify data access. The common gateway interface is a standard used by Web servers to communicate with applications on the server or on the network. It provides the necessary data access support that gives i-net applications their capability to pull data from data sources.

Just as the growth of the Web was enabled by underlying standards, relational database access has fundamental protocols creating an open environment. SQL is now the standard access to relational databases. And Open Database Connectivity (ODBC), despite early setbacks, has become the de facto standard in client/server database access.

The foundation for all this will be database management. Dynamic forms can collect information and store it in databases. Dynamic Web pages can query those databases and use the results to generate more Web pages on the fly.

7.5.8 Changes in Corporate Culture

Intranets do more than link people and data. Intranets promote cooperation and information sharing.

Intranets can affect a company's whole corporate culture. By making information readily available, an intranet changes the responsibilities of employees. Those with information can no longer sit back and wait for colleagues to ask them for it. With intranets, they need to make that information available in the public domain. Active information sharing is counter to most organizations' historical corporate culture—which is usually based on internal competition.

For global companies, the changes wrought by intranets are a necessity. Information sharing is a precondition to operating a global franchise, and companies that want to establish or maintain a global presence must eliminate internal competition and promote ways to share information.

7.6 Intranet Standard-Setting Bodies

Open User Recommended Solutions (OURS), a nonprofit industry group based in Cambridge, Massachusetts, and established in 1991, is dedicated to promoting open IT standards that can be used to link intranets to specific business functions. Members typically are *Fortune* 1000 companies. User-side members include Levi Strauss, Merrill Lynch, and Texaco, and vendor members include IBM, Microsoft, Novell, and Oracle. Besides intranet applications, other issues on the OURS agenda include information security, mobile computing, network management, and software licensing. The intranet task force is focused on the intranet's impact on users as opposed to any specific technical issues—to do a better job than when the first PC was introduced. OURS can be reached at www.ours.org.

7.7 Intranet Success Stories

What follows are just a few examples of organizations that have looked to intranets to improve their competitive advantage in the marketplace.

Social Security Administration

Even the Social Security Administration (SSA) is using an enterprisewide intranet as a key part of the overhaul of its business processes. Its goal is to speed policy development and cut associated costs. The intranet is the foundation for a workgroup-based organizational structure and will speed the process of disseminating and sharing information among the field offices, making decisions, and handling and propagating change throughout the organization.

The SSA expects to eliminate 5000 positions from its workforce by 1999, and in that same year, it expects to receive 84 million calls on its toll-free line and perform 864,000 continuing disability reviews—a substantial increase from the 62 million calls and 285,000 reviews SSA handled in 1995. In addition, a law passed in 1996 requires the SSA to automatically issue Personal Earnings and Benefits Statement forms annually to people aged 60 and over. By 1999, it has to expand this service to people aged 25 and over.

Revamping its business processes and leveraging an enterprisewide intranet will help the SSA deal with the growing workload despite workforce reductions. Access to information through the intranet will let an SSA employee handle an inquiry from start to finish, allow customers to deal with one employee throughout the course of their inquiries, and reduce the amount of time spent on each case, freeing up staff for other work.

Greyhound

Greyhound Lines has a corporatewide intranet designed to be a one-stop shop for everyone from part-time ticket agents looking for basic corporate information to the CEO, who gets a robust executive-information system. The intranet offers the company's army of employees and independent agents the latest in push technologies, synchronized back-end databases, and high-level data-querying tools—all through a simple, browser-based interface (in this case, Netscape Navigator) that can satisfy everyone's computing needs at a cheap, thin-client price.

The intranet project began as an offshoot of Greyhound's Internet site—a fairly sophisticated corporate site that provides customers with ticket pricing, bus stop and route locations, and other kinds of time-sensitive information. This Internet experience generated the in-house TCP/IP and Web-based interface expertise needed to take on such a large intranet project.

Pfizer

Pharmaceuticals giant Pfizer, Inc., relies on its intranet to distribute government-mandated information on occupational hazards in the workplace. Every chemical brought into each of Pfizer's 30 manufacturing sites worldwide must have hazard communications that comply with international and U.S. government regulations. Those Material Safety Data Sheets (MSDS)—some 200,000—run about five pages each and must be updated regularly and made available to all employees.

Pfizer built the Technical Information Exchange System (TIES) to update and disseminate the safety data, making MSDS available to all employees. The manual system required the MSDS to be sent to every site, where they were photocopied and then manually inserted into binders.

In addition, before TIES, Pfizer used to send out copies of pertinent sections of the daily announcements from the U.S. Occupational Safety and Health Administration (OSHA) and the U.S. Environmental Protection Agency (EPA). Now Pfizer finds the Environmental Health and Safety sections of the register on the EPA and OSHA sites on the Web and calls them to the attention of appropriate employees by posting links to the exact parts of the Federal Register that relate to OSHA, EPA, and international regulations.

Detroit Automakers

In Detroit, the Big Three automakers see intranets as a speedy way to gather information from dealers and to get quick responses through the supply chain. More than 1.1 million employees work for the three auto giants, and millions more draw paychecks up and down the industry's supply chain. Nearly every auto-industry worker could one day use Web applications. The automakers have proved that intranets and extranets can help car makers open the supply chain, streamline international communications, and save millions on desktop support.

The most immediate cost savings identified by auto-industry IT departments are provided by cheaper global collaboration and low-cost application deployment. Ford and GM see intranets as key to global expansion, while Chrysler's engineering department leads the way in conducting collaborative workgroups through secure Webs.

The quick-moving auto giants have suppliers and retailers scrambling. First- and second-tier parts and systems manufacturers have come under increased pressure to make themselves accessible via intranets and extranets. On the retail side, dealers are being tapped to provide upstream information back to headquarters.

The Big Three are looking at intranets to reduce the time it takes to launch a new car or truck. A few years ago, U.S. manufacturers would take up to five years to bring products to market, and Asian models were consistently beating them to showroom floors. The Web may take float time out of the production process.

Streamlined supply-chain communication is a top priority. Today, it can take an average of 30 days for EDI (electronic data interchange) transactions to get through the supply chain. Web communication will shorten that time frame.

Security is a big concern in Detroit. Ultimately, all of the auto manufacturers are looking for full integration of information systems, regardless of national boundaries. But because the Web is not considered fully secure, private digital satellite communication systems will remain key for delivering encrypted information overseas. Market data, pricing, and new design concepts must remain closely guarded trade secrets and are unlikely to be exposed on an intranet.

Arthur Andersen

Arthur Andersen & Co. replaced a 35,000-page three-ring binder with an intranet called KnowledgeSpace, now available to employees worldwide. Andersen took the content from each of its major businesses, which specialize in business practices, information technology, accounting, and vertical industries. Using KnowledgeSpace, individual consultants can bring the company's collective knowledge to bear on their clients' business problems by using the client's Internet connection to access Knowledge-Space while at the client site.

Leveraging corporate knowledge through an intranet has become a matter of staying competitive in the professional services industry. KPMG Peat Marwick, Coopers & Lybrand, and Booz Allen & Hamilton, Inc., have major initiatives to put browsers on every desktop and move corporate knowledge into databases connected to Web servers. The intranet is an integral part of these companies' strategy—the intranet is considered an investment, not a cost.

7.8   Extranets

The Internet has become a pivotal element in the evolution of client/server technologies and application development. Incorporating a

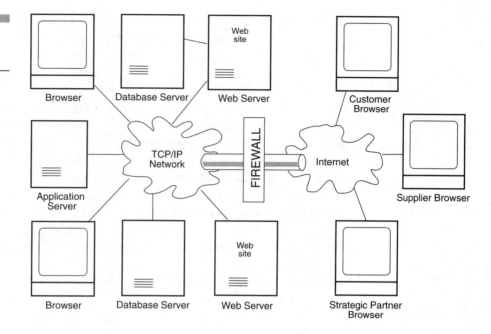

Figure 7-2
Components of an
Extranet

browser into existing client/server architectures has created powerful options for delivering information and applications. The evolution began with the connection of the internal DBMS to the Internet. It evolved into intranets and is beginning to evolve into extranets.

An extranet is part Internet, part intranet, as illustrated in Fig. 7-2. It allows companies to give trading partners—suppliers, customers, etc.— access to areas of their intranet as well as access to operational data via the Internet or a virtual public network. Federal Express has an excellent example of an extranet. FedEx Ship (downloadable from the FedEx home page at www.fedex.com) allows customers to track the location of their packages from anywhere in the world. Using the browser on their machine, customers sign on to the extranet, key in the parcel identification number, and are presented with detailed information on the location of that specific package.

Organizations that have incorporated EDI have had electronic links to suppliers. However, companies beginning to look at EDI are faced with high costs associated with setting up the infrastructure and the implementation and integration of proprietary software to customers and suppliers.

In contrast, implementing an extranet is relatively simple. It uses standard Internet components—a Web server, a browser, and possibly applet-based applications. The Internet itself can be used as a communications infrastructure. A firewall between the internal network and the Internet provides the security that allows only recognized users into the internal network. Regardless of their size, companies can use the extranet because it's accessible via an Internet link and requires no proprietary software.

Extranets can be used as a form of self-service—suppliers and customers can browse the available information, find what they need, and expedite business. When employees of General Electric and other companies with purchasing agreements with Dell Computers access the Dell home page, they see a customized page based on the buying guidelines laid down by their management. National Semiconductor's extranet allows design engineers at customer locations to download datasheets and order sample parts—a feedback loop pushes this activity back to National's marketing, engineering, and sales departments.

But it isn't quite as easy as it sounds. Once the extranet is implemented, many conditions are out of the control of the hosting organization. Desktop configurations may not be adequate at connecting sites. The hosting organization needs to "sell" its extranet to ensure its use. The hosting organization needs to be prepared for support calls from external users and for a variety of levels of proficiency.

PART 3

Deployment

Developing i-net applications requires a new approach to thinking of possibilities—truly to think "outside of the box." Key to the success of i-net applications is adherence to standards and protocols. Interestedly enough, there are more choices, rather than fewer. There are more decisions to be made, rather than fewer. And there is more visibility, rather than less.

An infrastructure must be in place to address security and reliability issues. Someone in the organization has to play devil's advocate: What if this happens? or this? or this? With information suddenly available to everyone internally, for example, data security becomes a very big issue.

Plans for managing the intranet and its applications must be in place. What are acceptable levels of service? How will the suddenly most precious of resources—the bandwidth—be managed? How can an organization provide reliability when most of the network nodes on the Internet are not under its control?

Organizations must begin to think, "What do I want to do?" rather than "How do I want to do it?"; "What do my customers want?" rather than "What will I give my customers?" This evolution isn't being fueled by companies wanting to reach their customers, it's being fueled by customers that want to reach companies.

Is it the chicken and the egg dilemma? Maybe. But ask yourself this: If I put up a Web site for my company because I wanted customers to come to it, how would I let them know it was there?

CHAPTER 8

I-Net as a Platform

United Parcel Service's Web site is a good example of how an i-net application works and what is done at each level of the architecture. At UPS's Web site, users can determine the status of a package they have sent, the transit time between two zip codes, and the price to ship a package given the package's weight, destination, and origin. The system first checks the Web user's credentials against a table of security metrics. Then a Hypertext Markup Language (HTML) form is sent to the user's browser from the company's Web server. The completed form is sent back to the Web server and run through a common gateway interface (CGI) script or some other program to convert it into SQL and trigger a query to a relational database. The results are formatted back into HTML, passed back to the Web server, and then forwarded to the user's Web browser. The user requires no special software and can be running on any platform, and there is no "on-hold" waiting. And the entire process takes only a matter of seconds.

Figure 8-1
I-Net Architecture's
Three Tiers

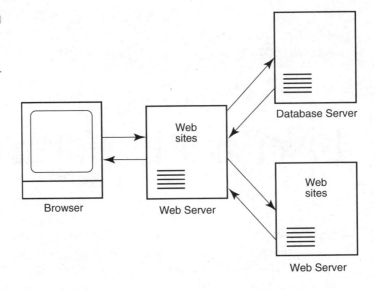

8.1   I-Net Architecture

I-net architecture is definitely segmented into three tiers, as illustrated in Fig. 8-1.

The thin client is on the first tier, with browser software and Java applets (small programs written in Java and downloaded with a HTML page) running as needed as the front-end application; the second tier is a Web server that provides HTML pages and Java applets and acts as a gateway to the third tier. The third tier consists of databases and other Web sites that may be accessed by the Java applet.

As technology continues to evolve, IT groups have more choices for these tiers. Fully loaded desktop PCs have long been the Net client machine. They are now being joined by network computers and Internet devices, as well as by inexpensive desktop systems with communications functions. Early i-net systems used Unix servers; they are now being joined by PC servers with built-in Web access.

8.2   Underlying Technologies

The components of an intranet and how they interact are illustrated in Fig. 8-2.

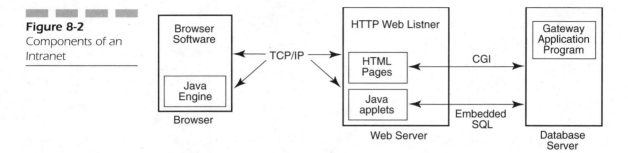

Figure 8-2
Components of an
Intranet

8.2.1 Hypertext

Hypertext provides the navigable links to related material and locations. Readers can use hypertext documents for linear reading or as a rapid search tool. Most on-line Help facilities are hypertext documents. This navigation is the basis for the forward and back concepts used on Web pages.

What hypertext does in an i-net architecture is remove barriers. There are no distinctions between internal and external documents or between Web servers. Users can reach any document on the Web from any other document on the Web, and once legacy applications become Web-accessible, that reach will also include legacy data and applications.

8.2.2 Hypertext Markup Language

HTML is a text-processing markup language that is used primarily for typesetting and hypertext applications. It is a subset of the Standard Generalized Markup Language (SGML), which is a text-based language for describing the content and structure of digital documents. HTML is used to structure Web page content.

The bulk of an HTML document is plain text. HTML tags, short commands surrounded by angle brackets, are placed around any text that is to be highlighted. For example, <b> and </b> would be the tags around text that is to be displayed in bold. Every hypertext page on the Web is a plain ASCII text document with embedded HTML commands that provide font and graphics information and links to other Web pages and Internet resources. Web browsers interpret these embedded commands to properly display the text and images. A sample of an HTML document with explanations is shown in Fig. 8-3.

Forms can be created using HTML tags. One of the arguments of the Form tag is an action attribute that is used to specify a uniform resource

Figure 8-3
Sample HTML with
Explanations

`<HTML>`	When used, these tags encapsulate the document, but they are optional, since Hypertext Transfer Protocol (HTTP) servers and browsers don't handle any other type of document.

`<HEAD>`
`<TITLE>` How HTML is formatted`</TITLE>`
`<META name = "keywords" value = "html, tag">`
`</HEAD>`
	A title is mandatory. It doesn't appear in the document but acts as a label in the browser's display area.

`<H1>`8.2.2 Hypertext Markup Language`</H1>`	H1, H2, etc., are heading levels. Heading tags render different font sizes as well as paragraph breaks before and after the heading.

`<BODY>`
 	
`<B>`HTML`</B>` is a text-processing markup language that is used primarily for typesetting and hypertext applications. It is a subset of the `<A>`HREF = "Exp-SGML.html">Standard Generalized Markup Language`</A>` `<I>`(SGML)`</I>`, which is a text-based language for describing the content and structure of digital documents. HTML is used to structure Web page content.
`</BODY>`
	The body of the document is indicated by the `<BODY>` and `</BODY>` tags.
	
	`<B>` and `</B>` tags indicate bold, and `<U>` and `</U>` tags indicate italic.
	
	If there were multiple paragraphs, each one would start with a `<P>` tag. A `</P>` is not usually needed, since the start of a new paragraph implies the end of the previous one.
	
	The `<A>` and `</A>` tags are anchor tags that encapsulate a section of text that is a hypertext link. The destination—in this case, Exp-SGML.html—is enclosed in quotation marks. The text between the anchor tags is rendered as underlined or in a different color from the surrounding text.

`</HTML>`	

locator (URL) to which to send the contents of the form. This URL usually specifies a script or program that can process the incoming data block. These programs use an application programming interface (API) called a common gateway interface, which is discussed below.

The Input tag is used to create text boxes, radio buttons, check boxes, and user-defined command buttons.

HTML tags provide in-line support for standard bitmap graphics formats such as .BMD, .PCX, and TIFF. HTML is extended by using GIF and JPEG graphics to add sound, full-motion video, 3-D, full-motion graphics, and interactive forms. Supporting these extensions are applets, which are little programs that only do one thing. The progressive GIF and JPEG formats, where a low-resolution version of the image is rendered almost immediately and then the image is progressively sharpened as the remaining information is received, are becoming more widely used. Adding in-line images requires adding an image tag, which is a link to an image file.

However, there are quite a few products on the market that will take care of generating the HTML codes from a form design created by the developer. These are called Web authoring tools. Some of the more popular products are PageMill from Adobe Systems, Inc., and FrontPage from Microsoft. Web authoring capabilities are also included in the most current releases of the major micro-based word-processing packages as well.

HTML as a specification has been slow to evolve even as Web browsers quickly adapt new tags. HTML's evolution is illustrated in Fig. 8-4. The draft specification for HTML 3.0, issued in March 1995, included support for such features as tables, background images, and mathematical equations. However, the differences between HTML 2.0 and 3.0 were so great that stan-

Figure 8-4
Evolution of HTML

	Features
HTML 1.0	Tags for formatting
HTML 2.0	Creation of forms
HTML 3.0	Creation of tables
HTML 3.2	Support for tables, applets, text flow around images, and superscripts and subscripts
HTML 4.0 (Cougar)	Standardization of object specifications, object tags, style sheets, and layout controls for doing layering and tiling of objects

dardization and deployment of the entire proposal proved unmanageable. The draft expired and is no longer being maintained. However, the standards proposed for HTML 3.0 are already in wide use on the Web.

The tags used by the current version of HTML specifications, HTML 3.2, add support for tables and are backward-compatible with any browser that supports tables, and they all use naming conventions that don't conflict with SGML specifications. The abiding principle of HTML 3.2 is to give Web authors greater control over layout through design elements such as tables and frames.

A significant addition in HTML 3.2 is Cascading Style Sheets, which let Web designers separate a page's layout and visual structure from its content, enabling more browsers to view at least the basic contents of a page.

W3C approved the new HTML 4.0 standard in January 1998. HTML 4.0, which incorporates many of the benefits of Dynamic HTML, is designed to create neutrality among competing browsers. It adds support for style sheets, and advanced forms and enhanced tables, and for objects and scripts. Its new functionality is not expected to appear in the marketplace until 1999.

8.2.3 Dynamic HTML

HTML produces flat, static pages. To make them interesting, organizations use HTML extensions, plug-ins, applets, animated GIFs, JavaScript, ActiveX controls, and VBScript. However, these solutions lack a universal standard.

Dynamic HTML is a new technology that will enable Web pages to have functionality similar to that of the desktop PC. The gist of dynamic HTML is its ability to change a Web page's format and layout without downloading an entirely new file. The images on the page will be movable at the user's request. Dynamic HTML promises to provide a basis from which users can run software applications without having to download and extract the program first. Not only does this save the user's time—there is no need to download, quit the browser, extract the program—it reduces network traffic because a server is not requested to download one file after another.

Dynamic HTML uses Cascading Style Sheets as a means of presenting documents on the Web using fonts, colors, etc., without changing the underlying HTML tags. It uses technologies such as the Document Object Model (DOM) that expose HTML attributes as properties that can be manipulated via scripts.

The World Wide Web Consortium (W3C) is working on setting the standards for Dynamic HTML. Both Netscape and Microsoft have submitted technical Dynamic HTML proposals to the World Wide Web Consortium.

Netscape's proposal uses a *layering* technique which sends browsers vast amounts of layered code that remains out of sight until needed. Netscape's version of Dynamic HTML is in its Communicator Web client. It does support Cascading Style Sheets.

Microsoft's proposal uses *cascading stylesheets* where photos or illustrations might appear as the mouse cursor is moved over certain parts of the screen. Its version of Dynamic HTML is the object technology in Internet Explorer 4.0 that makes HTML elements programmable.

Microsoft is also adding Dynamic HTML components to its development tools—ranging from Visual C++ to Visual Basic and Visual J++— to enable these tools to be used to create applications that can be accessed from any platform. Many industry analysts see this as a reaction to Java's growing momentum, but Microsoft's solution doesn't make the tools platform-independent. Visual Basic applications will still run only on a Windows platform.

Dynamic HTML is a key building block of Microsoft's new software architecture, Windows DNA (Distributed Internet Applications), which combines features from both Internet and client/server architectures to produce a framework for an application that can run over the Web or on a PC, laptop, or network computer. Windows DNA also includes COM+ (due out as beta in early 1998) for security and memory management as well as Object Linking and Embedding DB and Active Data Objects for easier data access.

The World Wide Web Consortium now believes that this approval process may take longer than originally expected because it involves a thorough look at all the issues and requires complete consensus among all members.

8.2.4 XML

The Extensible Markup Language (XML), a proposed specification from the World Wide Web Consortium, is one of a handful of HTML replacements, which also include HTML 4.0 and Dynamic HTML. XML will make documents multidimensional, capable of being processed by different programs, delivered by different methods, and viewed differently by different users.

The main problem is that HTML does not go far enough in automatically supporting customized markup and Web-page elements. Developers and vendors get around HTML's limitations by adding new page layout elements to HTML or by using Java, JavaScript, and special Web-enabled applications to make HTML look prettier.

Although it is still being defined, XML is generating excitement because it lets developers create custom tag sets for building cross-platform applications across the Web that are data-neutral yet more structured than what is possible today using straight HTML. By allowing users to define their own tags, the problems now caused by proprietary extensions to HTML will be eliminated.

As with HTML, XML document components are marked using tags, and documents created using both standards will be viewable using the next generation of Web browsers. However, unlike HTML, which defines a set of tags to describe data, XML lets users define their own tags and include the definitions within XML documents.

XML will enable applications that use Web browsers to integrate data from multiple databases with different formats. Electronic commerce on the Web, which has long been handicapped by the complexity of data formats, will also benefit. XML can effectively offload applications from overworked servers to desktop PCs. While it's now possible to do that using a Java applet, XML offers a more elegant approach that avoids the need to bundle the data with the applet.

XML is a very economical way to structure large amounts of unstructured data or data stored in document mangement systems. Such a feature will allow searching and information discovery applications to run faster and easier. For example, XML tags can serve to define searchable database fields.

Developers are interested in XML because many feel that Java is not yet mature enough for client-side business use. The XML style-sheet language lets Web creators define their own tags and requires more data-centric display capability than the document-centric HTML.

Other possible applications for XML include payment systems [Microsoft, CheckFree, and Intuit are developing an XML-based standard called Open Financial Exchange (OFX), which will handle payment instructions to banks] and push services. Microsoft's Channel Definition Format (CDF) is based on XML. The Meta Content Framework (MCF) has been proposed for designing visualization controls, push features, and more. A mathematical markup language (MathML) describes mathematical expression structure and control.

8.2.5   Common Gateway Interface

With the traditional approach to i-net applications, a very thin client communicates via HTTP to a Web server. The server's main purpose is to serve documents or files that can contain very sophisticated graphics that are files in themselves.

For its intended purpose, page description, HTML is fine. But for customized features, HTML is not much help. These features are implemented on the server as a common gateway interface. CGI is the standard that emerged to support data access and is used by Web server software to communicate with other applications on the server or on the network. It is a CGI script that gives Web servers capabilities such as forms processing, image mapping, and links to database software and other packages. CGI scripts are executed by the server in response to user actions.

For example, if a Web page contains a searchable product catalog, a user can enter a keyword into an HTML form. The CGI application sends the request to a database program and tells the database program to return the results of the search as an HTML page. This CGI application generates a new page each time the database is searched.

CGI is discussed in more detail in Sec. 10.7.1.

8.2.6 NSAPI, ISAPI, and ICAPI

An alternative to extending the abilities of the server is to use its API. The API allows the developer to modify the server's default behavior, which forces the server to use the code in an API module instead of its own built-in code. However, unlike CGI, API functions are server-specific because each server has a different API. Developing the code is more complex, since the API must manage multiple process threads and clean up memory after it is run.

Web server vendors have also created APIs for server-resident applications, which are designed to improve performance by eliminating the process-creation overhead of CGI. They do improve performance, but the application has to fit into the server's address space—an expensive proposition. In addition, these APIs are proprietary and do not specifically address database access.

Netscape API (NSAPI) allows a browser to execute programs on a Netscape Web Server. Using NSAPI function calls, Web pages can invoke applications on the server, typically to access data in a database.

Internet Server API (ISAPI), from Microsoft and Process Software Corp., allows Internet browsers that support ISAPI to access remote server applications set up on Microsoft Internet Information Servers (IIS). Using ISAPI calls, Web pages can invoke executable code on the Web server.

IBM's Internet Connection API (ICAPI) provides control over IBM's Internet Connection Server product. It is basically a clone of NSAPI that runs on IBM's server products.

These APIs are discussed in more detail in Sec. 10.7.2.

8.3   The Network

Most intranets inside corporation walls run at a minimum of 10 Mbps (million bits per second). Most Internet connections use a T1 line (1 Mbps), so the inside pipe is 10 times bigger than the outside pipe. Consequently, bit-rich data types such as video and real-time audio are real options for intranets because they stay "inside."

8.3.1   Transport Protocols

IP, IPX, and NetBIOS are the primary transport protocols on today's networks. IP is the protocol used by the Internet, and IPX is the most common on LANs. An IP/IPX bridge is used to connect an intranet on IPX to a Web server using IP. NetBIOS bridges exist but aren't as common.

8.3.2   IP Assignments

An intranet platform should also have a Dynamic Host Configuration Protocol (DHCP) server to set up and manage static IP assignments. Windows NT Server comes bundled with a DHCP server. Other Web servers are supported by third-party products.

A domain naming system (DNS) is a mechanism used in the Internet for translating names of host computers into addresses. A single DNS server can track multiple Web servers on the intranet but needs to get dynamic IP addresses as they are assigned via DHCP. Usually the DNS and DHCP packages come from the same vendor, so as DHCP addresses are assigned, the names are also logged into the DNS database.

8.3.3   Bandwidth

The size of the pipe (the network's bandwidth) dictates the capacity of the network. It is measured in cycles per second—bits per second (bps) for digital channels and hertz for analog channels.

The amount of bandwidth a channel can carry dictates what kinds of communications can be carried on it. A wideband circuit, for example, can carry one video channel or 1200 voice channels.

Bandwidth compression is a technique used to reduce the bandwidth needed for a particular transmission. It is usually used in image transmissions such as fax, imaging, or videoconferencing. Bandwidth compression can be used for voice and video as well as data. Any compression done at the sending node must be decompressed at the receiving node, and the routines must be compatible.

Bandwidth allows a network to ask for (and ideally get) additional circuits for a short period of time. Bandwidth on demand is provided only with digital circuits because they are easier to combine. Bandwidth on demand is usually carved out of an existing T1 circuit.

Fixed-bandwidth services provide point-to-point links at constant rates. T1 consists of twenty-four 64-kbps [kilobits (one thousand bits) per second] channels and can transmit at rates up to 1.544 Mbps. T3 combines 28 T1 lines, giving it a transmission capacity of 44 Mbps. T3 requires fiber-optic cabling and can carry 672 voice or data channels at 64 kbps.

8.4   Development Languages

Currently application logic is coded into HTML pages using ActiveX controls or Java code. Java code is downloaded with an HTML page to the client and discarded when the page is discarded. ActiveX controls are downloaded only if they are not currently installed on the client. They remain installed on the client machine until they expire or are manually deleted.

ActiveX and Java are discussed in more detail in Chap. 13.

8.4.1   Java

Java executable programs are called applets. The applet process is illustrated in Fig. 8-5. Applets are downloaded with an HTML page to the client. The applet is interpreted on the client and executed by the Java Virtual Machine (Java VM) environment. The Java VM enforces the executable memory boundaries, performs automatic garbage collection, and prevents the applets from writing on the local file system. However, inter-

Figure 8-5
Java Applet Process

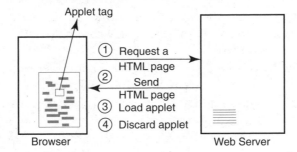

preting code one line at a time slows down execution. Just-in-time Java compilers that will speed up the execution time are now appearing on the market. The applet stays in memory until the client machine is turned off, at which time the applet is discarded.

The Java language coupled with HTML for layout and graphics allows dynamic applications to be downloaded and run as needed on the client. The client can use applets to do animation, generate a form, perform calculations, do data massaging, or generate local reports.

Applets can run only on browsers equipped with Hot Java. Hot Java acts as an interpreter to let interactive Java-based programs run on the users' workstations regardless of platform. Applets run within a Web browser regardless of whether the browser is running on a Windows 95 micro, a Unix workstation, or a Power Macintosh. The developers of Java call this "write once, deploy anywhere."

Java is being used to perform field edits locally on the client, to reduce the number of network interactions. A Java database can be accessed using Java Database Connectivity (JDBC) drivers that provide the specification and API to make SQL requests. The drivers are downloaded as Java-class files over the network, just as applets are. The JDBC Driver Manager chooses which low-level driver will handle the target DBMS request.

Intersolv's JDBC-ODBC (Open Database Connectivity) bridge communicates with existing ODBC drivers to make database calls. This is better suited for intranet environments because the bridge must be installed on each client.

8.4.2   ActiveX

The ActiveX approach integrates component libraries at the desktop with new controls that are loaded dynamically. If a control needed for a Web page is not already resident on the desktop, it is brought over with the Web page. The advantage is that run-time performance improves (if the control is already resident, it isn't downloaded). The disadvantages are the need for disk space for the component libraries and the reliance on Microsoft environments. To respond to this criticism, Microsoft recently released a software development kit that helps port ActiveX controls to Macintosh and Unix.

ActiveX components plug into the container that controls them—the browser, for instance. ActiveX is different from Java in that the loaded controls are stored in regular directories at the client for subsequent use. If the controls are missing from the desktop, they are located using the <OBJECT> tag, which points to the servers that store them.

ActiveX does have scripting languages—VBScript and J-Script. Java can also interact with the ActiveX environment using tools such as Microsoft's J++ visual development tool.

8.5   Extranet Architecture

For an extranet to support outsiders, they need to be able to get into the intranet—securely. One configuration would be a direct leased line from intranet to intranet. Another is a secure link over the Internet. In this scenario, it makes sense to have servers outside the firewall communicate with servers inside the firewall. (A firewall provides a boundary to prevent traffic from one segment of a network, such as the Internet, from crossing over to another segment, such as an intranet, without authorization. For more information on firewalls, see Sec. 10.4.)

Security is the key. There will be areas of the intranet that outsiders can access and areas that they can't. Directories will have lists of entries for people in the company and lists of entries for people at suppliers. These people literally come into the network (via the Internet) and log into the directory. They authenticate themselves to the directory just as internal users do, but have access to only a subset of resources.

8.5.1   Extranet and Data Access

The seamless integration of data between the browser and server is key to the success of an extranet. This capability is an evolution of client/server technology. For example, Federal Express links its Web site directly to its operational systems using a relational DBMS and middleware technology. From a technological point of view, such database-enabled Web applications are just another form of distributed client/server applications. The information accessible by the extranet relies on the middleware tools that connect the dispersed systems to the DBMS source as well as the Internet browser.

Early Web sites provided one-to-one marketing. Every visitor was provided with the same views and information. By integrating Web and database servers, organizations can maintain and store profiles on users that allow them to tailor information to the user's interests. This information can be captured at an initial visit and used to provide custom information and promotions on subsequent visits. For example, if on the last visit to amazon.com, a user looked only at mysteries, on the next visit, the user might see a flash item on a newly released mystery as well as other books by that same author.

8.5.2   Software for Extranets

Netscape is leading the charge toward extranets with a line of products optimized for extranets. The products are referred to by their code names.

- *Mercury,* the next version of the Communicator client suite, will have an object store so that users can download applications from a network, store them locally, and use them off-line.

- *Navigator,* the browser component of the suite, will have a new rendering engine (code-named Gemini) for faster rendering of HTML, Java, and JavaScript.

- *Hypertree,* a file management system, allows users to organize E-mail, local files, and network files in a single window. A new component called Compass will allow users to organize files in a personal HTML page stored on a server.

- *Apollo* is the next version of Netscape's SuiteSpot suite of servers. It will have an object store and will include services that support workflow and transaction-processing applications built from distributed objects.

- *Palomar* is a visual development environment for building cross-platform applications from HTML, Java, and JavaScript components. Using component interface technologies such as JavaBeans, developers will be able to access native component models such as OpenDoc and ActiveX.

Palomar was released in late 1997. The others, as well as a second version of Palomar, are expected in early 1998. For more information on these products, see Sec. 19.5.2, "Crossware."

CHAPTER 9

Clients and Servers

As organizations began to implement client/server archi-tectures, they began to think in terms of clients and servers and the division of duties. This mindset is required for successful i-net application implementations.

9.1   Thin Client

The term *thin client* is used to strongly differentiate the client used in Web-based applications from the client used in client/server applications. When executing applications, a client/server client machine runs software that is stored locally. Some of the application may be stored and executed on a server, but most of it is on the client. The server provides the data for the application.

In Web-based applications, the server provides execution code as well as data to the client as needed. The browser software is the users' interface. Consequently, the only software the client needs, in this case, is an operating system, browser software, and software to convert downloaded code.

The original idea of a thin client was a machine without a floppy or hard disk—basically a terminal with smarts. The hardware definition of a thin client is changing, but the concept behind a thin client has not—get the code and data needed when needed, and discard after use.

Some of the benefits of these server-run machines include remote administration of software and configurations and security. There is centralized configuration management as well as the administration for deployment and maintenance of software. Data is secure at a central site. Users can't walk away with corporate data on a storage medium. Unlike laptops, these client machines aren't stolen, since they are almost worthless without a network connection.

9.1.1   Network Computers

Network computers (NCs) are being marketed as the perfect thin client hardware. These machines have no floppy or hard disk storage and sell for under $400. They use a compact operating system that can be booted from the network and include a Web browser and Java Virtual Machine (Java VM) for running Java applications. They may also include a smart card slot for user login verification. They are designed to run applications that are acquired (downloaded) from somewhere on the network. In most cases, network computers run Java applets within a Java browser or Java applications.

NCs are designed for business use and can hook up to standard computer monitors. Web boxes such as those from WebTV (which was recently purchased by Microsoft) that connect to a TV set are aimed at home users.

Apple, IBM, Oracle, Netscape, and Sun were among the members of the consortium responsible for the creation of the Network Computer Reference Profile, which outlined the basic system design and open standards supported through the network computer. These include Java and Hypertext Markup Language (HTML) programming languages, and other Internet and networking protocols such as TCP/IP.

At the behest of the consortium, The Open Group was asked to develop a more standardized Network Computer Reference Profile in an attempt to reduce confusion in the NC marketplace. Vendors can license the specification and build their machines in compliance with it. The Open Group will help standardize the NC by testing potential NC devices for compliance with the profile and awarding them an official logo.

In 1996, Oracle created a subsidiary called Network Computer, Inc. (NCI), for marketing network computers, and trademarked NC at the same time. In mid-1997, Netscape and Oracle merged their respective companies—NCI and Navio Communications—into one company under the NCI name. Among the software under the new NCI umbrella is the Navio product family. These products are based on Netscape Navigator and are designed to bring Internet technology to both the consumer and non-PC markets. NCI software products are designed to build and deploy network computer systems suitable for homes, schools, and corporations.

Sun offers the Java Station as a network computer. The Java Station runs the Java operating system and Java-based applications. It features a Hot-Java Web browser and can access Windows applications. The Java Station architecture stores users' computing desktop parameters on a server, so that their desktops are available from any machine.

The IBM Network Station family offers solutions to targeted business needs. The IBM Network Station Series 100 can be considered an "access network computer" because users see an easy-to-use graphical user interface (GUI) connecting to applications residing on multiple servers. The Series 100 gives users access to popular business productivity tools, such as Lotus SmartSuite and Windows applications. The Series 100 is a logical replacement for terminals or PCs that are primarily used for accessing applications that reside on various servers, such as a travel agent booking airline, car, and restaurant reservations.

The IBM Network Station Series 300 is IBM's network computer for the Internet. It is designed for business users who require access to multiple servers, as well as intensive use of applications and data residing on corporate intranets or the Internet.

The IBM Network Station Series 1000 is IBM's "Java network computer." The Series 1000 is designed for businesses that are planning Java-based applications, such as those being developed by Lotus Development Corp., and need robust support for Internet standards.

9.1.2   NetPC

The NetPC is a Windows-based network computer from Microsoft and Intel. The NetPC is a combination network computer and slimmed-down Windows PC designed for standardized environments that require less flexibility. It is actually a full PC running Windows and Windows applications from a local disk, but, like the NC, it has no slots and comes in a sealed case—a fixed configuration that a user can't change. But it can be told to "wake up" from a network and can be managed remotely. It sells for about $1000.

The NetPC supports Java applications through the Java VM built into a Web browser. A NetPC does not need a network to boot because of its standalone architecture. A live network connection is needed only when a user wants to browse or load a network resource.

However, users can't install software on a NetPC. It has to be installed by a server. So if a user wants Microsoft Word on a NetPC to use when the server is down, the server would first have to download a copy—any idea how long that would take?

The benefit of remote administration will become diluted when Microsoft includes "Zero-Administration" Windows (ZAW) software in Windows 98 and Windows NT 5.0. ZAW will support server-based management of clients' hard disks. Each user's data and configuration information would be automatically copied to a central server. Frequently used applications would be cached on users' disks but automatically updated. By logging into any NetPC, users would have access to all their programs and data.

In addition, IT managers are asking why they should buy a PC that can't be upgraded—when it becomes obsolete, it has to be replaced. Consequently, acceptance of NetPCs has been slow. NetPCs are available from companies such as Dell, Gateway, Compaq, Hewlett-Packard, and IBM.

9.1.3   The Right Client for the Job

NetPCs and NCs are forcing organizations to step back and decide how a machine is going to be used. As illustrated in Fig. 9-1, the different machines actually implement different architectures. An NC from Oracle or IBM's Series 100 or 300 expects an open environment. Sun's Java Workstation expects all applications to be Java-based. A NetPC client is actually running Windows and supports Java through a Java Virtual Machine built into the browser.

Organizations need to think about how the machine is to support the

Figure 9-1
Thin Client Architectures

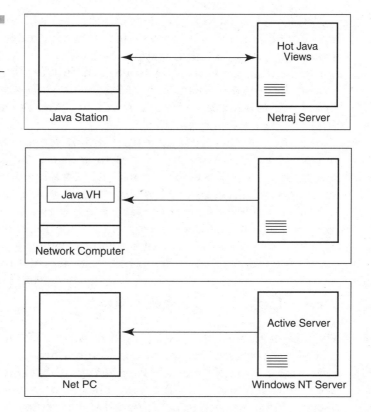

worker. A knowledge worker who uses a variety of tools and takes work home needs a hard drive and a floppy disk, and can't be confined to what's available via the network—or wait for it to download! This worker is using spreadsheet, presentation, database, and word-processing software as well as specialized programs. The job processes are not clearly defined, and functions are under the individual's control.

However, the user (workstation) who runs a couple of customized line-of-business applications such as customer service or order entry plus maybe E-mail doesn't care how the computer is configured. And yet these are the machines that cause the headaches for network managers—keeping these hundreds of identical systems identical.

The network computer is, after all, a computer. It receives Java byte code which allows the network computer to perform real local processing of applications, although some applications might run on the server. Either way, with processing power on both sides, there should be high performance with minimal bandwidth consumed. But Java is still a young product, and products written using Java are far from tried-and-true.

NetPCs appeal is to network administrators who are used to supporting PCs and who are not interested in deploying a new platform. If a PC is replaced by an NC, all the applications that run under Windows will have to be changed. Currently, an NC can only log into the network and browse the Web. Because a Java VM is included, Java applications can be downloaded from the server and stored in RAM. But an NC cannot run Windows-based applications.

Another point to consider is that hard drives do more than store data and programs. They are used for caching data from RAM. An NC will crash if an application won't fit into RAM. So without a hard drive, an NC needs enough physical RAM to run the Java VM, a Java-based operating system, a Web browser, and the largest application or set of applications likely to run on that station. In addition, local caching is used by Web browsers to store a local copy of the most frequently accessed pages. This copy is used rather than downloading another copy from the Web.

9.1.4　Cost of Ownership

The idea of thin clients has captured IT managers' attention, not because they are relatively inexpensive to purchase, but because their cost of ownership is a lot lower. Systems are kept identical because the software comes from the server as needed—there aren't hundreds of copies to update. When operating systems need to be upgraded, the server sends down a new copy.

Cost of ownership falls into five categories:

- *Capital* includes hardware and software such as the operating system.
- *Technical support* includes help desk, documentation, application consulting, user groups, learning, and disk management.
- *Administration* on the desktop side includes asset management, security, and auditing.
- *Network* includes capacity; moves, adds, and changes; upgrades; server purchases; and network operating system administration.
- *End-user operations* include data management, application development, and learning.

Considering these five categories, a Gartner Group study in early 1997 concluded that companies could achieve the following cost savings over Windows 95 desktop environments: 26 percent for the NetPC and 39 percent for the NCs. But temper those statistics with the fact that the study also states that it is difficult to conclude that NCs will be significantly cheaper because they are new and have not been implemented in sufficient numbers yet. In addition, the $400 price quickly grows to $1000 as more memory is added as well as a monitor.

One thing to remember about NCs is that the servers they are connected to have to be pretty powerful. They must be able to provide the high performance and reliability that users need so as to not leave them stranded without server or network connections.

And with NCs or NetPCs, the bandwidth has to be able to handle the traffic volume. Organizations using network computers will have to beef up their LAN and WAN infrastructure—those costs alone would easily wipe out anticipated savings associated with buying these lower-cost boxes.

9.2   Browser Software

A Web browser is software that allows a computer user to navigate— surf—the web of interconnected documents on the World Wide Web. Every time a user visits a Web page, the Web browser moves a copy of the document on the Web to the user's computer using Hypertext Transfer Protocol (HTTP).

In order to view a site, a user types in a uniform resource location (URL)—the site's address—and that page is downloaded. Links from that site may take the user to other related sites.

Mosaic was the browser that started the Web revolution; Netscape's Navigator (and its newer product Communicator) and Microsoft's Internet Explorer now vie for the top spot.

It is interesting to note that despite all the press about thin clients, these browsers are far from thin. Internet Explorer 4.0 needs 53 Mbytes; Netscape plans to unbundle the browser component in Communicator, but until that happens, a standard version of Communicator requires 30 Mbytes. In contrast, a standard installation of Windows 95 is a mere 40 Mbytes.

To keep disk space at a minimum for thin client platforms, many organizations are paring down the browser versions they install on those platforms. Internet Explorer 3.0 takes about 14 Mbytes for a typical version— Arthur Andersen & Co. gave its 55,000 users a 3-Mbyte version.

9.2.1   Common Browser Features

Today's browsers have come a long way from the functionality of even Netscape Navigator. They are now Internet suites which deliver tools for all on-line tasks, with support for mail, security, authoring, collaboration,

Figure 9-2
Browser Components

Text & Graphics	Plug-ins	Containers for components	
HTML	Shockwave	VB Script	Java Script
	Acrobat		
	Real Audio		
	Cool Fusion	Active X	Java Applet
	Live 3D		

and push content—and, of course, Internet browsing. The browsing portion of these suites is illustrated in Fig. 9-2.

Plug-ins provide the interactive capabilities that HTML doesn't. Plug-in technologies execute on the desktop while the user is in the browser. The executable code for the plug-in resides on the server and is downloaded to the client when necessary. Plug-ins perform one function for specific file types and are platform-specific. Those shown in the figure are some of the more common plug-ins used today.

Plug-ins do have their drawbacks. They download "just in time" when needed, forcing the user to wait while that download happens. Plug-ins are also proprietary in nature, so they are likely to be replaced with Java and ActiveX code in the near future.

The major names in browser technology today are Netscape Communicator and Microsoft Internet Explorer 4.0. They both have Network News Transfer Protocol (NNTP) newsreaders and are POP3- and IMAP4-compliant, HTML-aware E-mail clients. Both tools have Web-page editors, push delivery systems, and live-collaboration tools. Each has lightweight installations which omit many of these components, if desired.

Both have extensive drag-and-drop support and control over attributes such as security. Both support automatic software updates. With both products, reading NNTP news is just like reading E-mail.

Both browsers support basic Level 1 Cascading Style Sheets and Dynamic HTML. However, the models used in the two browsers are incompatible.

Both tools offer audioconferencing, whiteboard, text-based chat, and file transfer. Netscape has a two-party, point-to-point architecture, while Microsoft has a multipoint architecture, allowing three or more users to work together, although only two can take part in audio- and videoconferencing at a time. During chat, file transfer, and whiteboard conferences, Netscape can manage only two conferees, whereas Microsoft can support multiple users.

Both tools include a Java Virtual Machine and support for the Java

Development Kit 1.1, which includes the new Abstract Windowing Toolkit (AWT) and JavaBeans.

The major differences between these two products is the way they view the market itself. Internet Explorer 4.0 merges the Web browser and the operating system, becoming a browser for both local and Web content. The desktop wallpaper becomes an "active desktop" that can host miniature Web pages.

Netscape sees its browser suite as the client side of an enterprisewide, cross-platform architecture whose goal is the sharing of information both within a company and outside it. Netscape ships products for 15 platforms; Internet Explorer 4.0 is available only for Windows 95 and Windows NT.

Most organizations will end up supporting both. Web sites use one or the other as their authoring tool, and an individual site will "look and run" better if the client uses that authoring tool. As long as both products remain relatively inexpensive (or free), users will keep both on their desktops and use whichever one is appropriate for their Web surfing needs.

9.2.2 Microsoft Internet Explorer 4.0

Internet Explorer has been upgraded, with better control over viewing and navigation, and enhanced to include support for push channels and off-line browsing.

Integrated with Internet Explorer's browsing tool are

- *Outlook Express,* for E-mail (POP3 and IMAP4 can be routed to a single box) and news offerings
- *FrontPage Express,* for HTML authoring (a subset of the Microsoft product FrontPage)
- *NetMeeting,* for collaboration

9.2.3 Netscape Communicator 4.0

This suite of products consists of

- *Navigator 4.0,* an upgraded version of Navigator that includes a drag-and-drop toolbar, a complete history list, and a better download feature
- *Netcaster,* for support of push channels (Webtops are channels that are locked onto the desktop) and off-line browsing
- *Messenger,* for E-mail

- *Collabra,* for news offerings
- *Composer,* for HTML authoring
- *Calendar,* for personal or group scheduling

Netscape has announced its intention to release a Java version of its Communicator client in 1998.

9.3 Web Servers

A Web server is a computer that contains the documents and files that are displayed when people access the server via HTTP. The term is also used to include the software that manages the Web networking functions in a Web server: accepts requests from Web browsers and transmits HTML pages and other stored files. Consequently, buying a Web server now means buying the hardware—optimized for Web applications—and software for managing and hosting Web applications. As illustrated in Fig. 9-3, Web servers connect with the services, data, and applications within the network using open standards and built-in methods.

A Web site is a set of Web pages that can be visited by Web browsers. A Web server can host more than one Web site.

In many three-tier architectures, the Web server shares the middle tier with an application server that holds the business and application logic

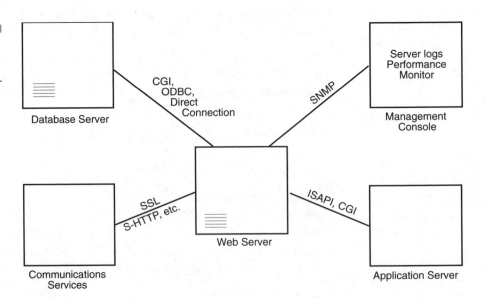

Figure 9-3
Web Server
Integration

necessary to support complex Web transactions. What separates the high-end and low-end servers (and their software) is how they handle transactions and how many they can handle.

Both intranet and Internet Web servers are IP-equipped servers that supply HTML documents. Products for these platforms range from simple HTML delivery to highly integrated database Web servers. An advantage in the intranet world is control over Web browsers, which deals more with content than with the platform itself.

The main consideration when choosing a Web server is the operating system. The major operating systems used by Web servers today are Unix, Windows NT, and NetWare. However, there are a lot of other options for first-time implementers—Windows 95, OS/2 Warp, and Linux, to name a few.

The Web grew up on Unix, which remains the premier platform for large Internet servers and large intranets. Unix is robust and a proven entity, making it an ideal platform for the largest number of Web server products on the market—although Windows NT is catching up quickly. Unix offers high-performance multiprocessing and clustering support features. Internet service providers like Unix because it can support multiple Web sites on a single system with a growth path to a multiprocessor system as the site—or the number of sites—grows.

If the server has more horsepower and more memory than a workstation, it pays to run Windows 95 on the server. Windows 95 provides 32-bit multitasking application support, which is critical to a Web server's performance. Although Windows 95 lacks the power and sophistication of other 32-bit operating systems like Unix, Windows NT, and OS/2 Warp, its low cost, low resource requirements, and 32-bit support make it an ideal platform for small to medium intranets. Windows 95 is also a good choice if all the workstations in the network run Windows or Windows 95.

Windows NT is a robust, 32-bit operating system with impressive security and network credentials. It is beginning to show up on most vendors' lists of supported platforms. Microsoft bundles its Internet Information Server (IIS) Web server with Windows NT. Windows NT's similarity to Windows 95 in programming and user interface is an additional plus.

Windows NT comes in two flavors: Workstation and Server. Windows NT Workstation is a great intranet platform, but Microsoft's latest licensing scheme might deter many from choosing it over Windows NT Server. Both versions come with a copy of IIS.

NetWare is still the file and print server used in the majority of networks today. Novell is trying to update its image with its latest version of NetWare, now called IntranetWare 4.11. Like Windows NT, IntranetWare comes bundled with its own Web server. IntranetWare's multiprocessor support makes it a good intranet platform for larger networks. Novell's NetWare Directory Service (NDS) also holds promise for securely managing large numbers of intranet users.

9.3.1 Intranet Web Servers

Intranet projects typically start small and grow. A 200-MHz Pentium Pro with 128 Mbytes of RAM might seem like a hefty server, but in many environments it is a small system. Large networks are just one place where multiprocessing systems come into play. Multiprocessing platforms are effective at improving file server and application server performance. They are equally effective for intranet servers, especially where Web servers are linked with server-based applications or server-based database systems.

The two major Web server software vendors are Microsoft and Netscape. Microsoft gives its Internet Information Server (IIS) away free with the purchase of Windows NT. Netscape offers Enterprise Server 3.5—but not for free. One of its major features is that it runs on NT as well as on most flavors of Unix and Alpha. Microsoft IIS runs only on NT. Another bonus for Enterprise Server, especially for smaller sites, is that it runs on Windows NT Workstation; IIS doesn't. But the most amazing part of Netscape's multiplatform support is that the versions for the different platforms look and act exactly the same.

New to the market is Lotus Go Webserver from Lotus. It lacks the E-mail and collaboration capabilities of Domino and doesn't require as much administration. Go Webserver is aimed at companies that are just starting out with Web development. Go Webserver is a rebranded version of IBM's Internet Connection Server. Lotus hopes that organizations that begin with Go Webserver will later upgrade to Domino or one of IBM's collaboration servers.

9.3.2 Web Application Servers

A Web application server contains application logic and sits between the Web server and the database engine. It manages the database connection on behalf of the browser. It compensates for the fact that the Web is stateless—that the Web server forgets the first interaction with a database even if the same client comes back nanoseconds later.

The common way to handle connects is to have a browser open a common gateway interface (CGI) scripting application on the Web server. The Web server launches an interpreter for the application and then it launches the application, accesses the database engine, acquires the result, shuts down the application, and closes the interpreter. The Web server then populates a Web page with the results and downloads the page to the browser. It goes through the same process the next time a client's browser—even the same client—asks for data from that same database engine.

An application server opens one or more connections with the database engine, stamps each user session with a unique identifier, and returns the results to the user's browser after interactions with the database. As the query load builds, the application server provides flexibility and scalability; it can initiate multiple copies of itself on one server or across multiple servers.

In some cases, an application server is in addition to an existing Web server. In others, they may be one and the same. Major database software vendors, such as Oracle with its Web Application Server, are adding application serving capabilities to their Web servers.

However, there are benefits from an application server's possessing its own hardware and memory resources. The database server can respond to queries better when it doesn't have to also manage user sessions. This also promotes easier scalability of the Web server and the database server.

However, application servers can be as expensive as the database engines themselves. In addition, they are complicated to set up. They are offered by newcomers such as Bluestone Software, NetDynamics, and Kiva Software. Familiar names such as Microsoft, Oracle, and Sybase are trying to put the application logic in the database management system itself.

9.3.3 Transaction-based Web Servers

Many database vendors are improving their connectivity middleware that links Web servers and databases. Vendors such as Oracle, Sybase, and Microsoft are providing development tools and object request brokers (ORBs) that keep the database transaction and state management functionality in the Web server and the database engine and out of the application server.

Microsoft offers Microsoft Transaction Server (MTS), which combines a transaction-processing monitor, object request broker (ORB), and application-development environment. MTS supports hundreds, rather than thousands, of users, but Microsoft is pushing ease of use rather than industrial strength. MTS will become more useful for enterprisewide applications as products that scale better with Windows NT, such as Microsoft's Cluster Server (formerly Wolfpack clustering software) and Message Queue Server, which were released in late 1997, begin to hit the market.

Sybase offers the Jaguar Component Transaction Server as its middle-tier connection. Jaguar allows developers to use standard components, including JavaBeans, ActiveX, and C/C++ components, and CORBA objects, to build and deploy applications that manage on-line Web transactions. CORBA stands for the Common Object Request Broker Architecture—an object request broker standard developed by the Object Management Group. CORBA provides for standard object-oriented interfaces between ORBs, as well as to external applications and application platforms. Jaguar was released in late 1997.

Oracle's Web Application Server is a Web request broker and a Web server that work with specific application cartridges. An application cartridge is a CORBA object that encapsulates the data and methods required for specific transactions. As with Sybase and Netscape, a key piece of Web Application Server 3.0 is CORBA ORB middleware licensed from Visigenic Software. (Netscape's object-based application server, Apollo, is expected in early 1998.)

Newcomer Kiva Software offers the Kiva Enterprise Server, which includes dynamic load balancing, transaction management, and multi-threaded and multiprocessing management. Developers can use the Kiva software developer's kit, Java, Visual Basic, C, or C++ to add application logic to the Windows NT or Unix server. On the front end, Kiva's server supports standard browsers, and Java or ActiveX applications. On the back end, the server can talk to most databases, including MVS data engines and transaction-processing monitors such as IBM's CICS. The focus is on enterprises with 100,000 or more users. Kiva has caught the attention of transaction-rich telecommunications companies.

In contrast, companies such as IBM and Informix are putting the logic into the core database engine. IBM's Universal Database Strategy pushes the middle tier into the database by storing the logic in the database as objects that can communicate efficiently with the database. Developers write DB2 Extenders (objects) that encapsulate new data types and associated methods, which then become part of the database server. The encapsulated methods contain the application logic to perform complex Web transactions. Informix is taking the same approach with its object-extension technology for the Informix Universal Server called DataBlades.

Critics of this approach say that pushing application logic into the core database engine makes the database less secure. IBM and Informix maintain that their approach is stable and results in better performance through tighter integration with a central database server.

It should be noted, however, that IBM has its feet in both camps. As part of its Network Computing Framework, IBM's Notes Domino server acts as both a Web server and a database that can hook into back-end CICS databases.

9.3.4 Bundled Web Server Software

Most organizations are looking for integration of all of these complex services. Currently, four Web server products provide this integration—out-of-the-box implementation.

Netscape SuiteSpot   Netscape SuiteSpot includes an Enterprise (Web) Server, a Messaging Server, a Proxy Server, a (multi)Media Server, a Calendar Server, the Collabra Server (a discussion group server), a Catalog Server, a public key Certificate Server, and a Directory Server built on Lightweight Directory Access Protocol (LDAP). SuiteSpot can be deployed on one or more servers. SuiteSpot also includes Netscape's LiveWire Pro, which allows developers and managers to manage complex Web sites.

See Sec. 4.2.3 for more information on Netscape SuiteSpot.

Oracle Web Server   Oracle's Web Server is a database Web server found on many large Internet Web sites delivering information from large databases. Organizations that already use a matching database server will find a product like Oracle's Web Server to be an excellent intranet tool.

Oracle's WebServer is a full-featured Web server with Oracle 7 access. It offers HTML-based Web server management, allowing any Web browser to act as a management workstation. WebServer is not completely integrated with Oracle 7. However, its tight integration and extensible database support make it an ideal platform for Web sites that support thousands of users and sites that require robust, transaction-oriented support.

Microsoft Commerce Server   Although it might not seem like an intranet product, Microsoft's Commerce Server is one, and it is a good choice for a large company. Commerce Server provides an effective way to show a variety of products as well as costs. Software distribution is one way Commerce Server can help within a corporation. Using Commerce Server, network administrators can post descriptions of available software packages, deliver them on-line, and automatically take care of the internal billing. In addition to Commerce Server, Microsoft offers other intranet servers and services to augment its IIS.

Lotus Domino   Lotus Notes' Web server is called Domino. Eventually, Domino will be a standalone Web server as well. The integration of the

Notes database and the Domino Web server is extensive, to the point that Notes' field-level security is present when Domino delivers an HTML version of a Notes document. Domino can provide access to any Notes database including mail, making it one of the most interesting Web environments currently available (especially given the extensive third-party support for Notes).

See Sec. 4.2.1 for more information on Lotus Domino.

10

Infrastructure Requirements

Implementing i-net applications requires an infrastructure that is focused on network communication—between clients and Web servers, Web servers and Web servers, and Web servers and database servers—and security.

10.1   IP Address Management

TCP/IP is the network of choice today: It is open, standards-based, robust, and WAN-efficient, and it is the protocol of the Internet. However, TCP/IP is an administrative nightmare. Each device on the network must have a valid IP address, and each IP address must be unique.

All devices on a segment must have the same network number and subnet number, and each subnet number must be unique. Communication with nodes on a different subnet or a different network requires a router.

The administrative nightmare? Assigned addresses must be tracked to avoid duplication. This sounds easy. For example, if a node is moved to a different subnet, the subnet portion of the new address must be changed to that of the new subnet. The resulting address must be checked to ensure that it doesn't conflict with the address of any other node on the new subnet.

Reverse Address Resolution Protocol (RARP) and the Bootstrap Protocol (BOOTP) were developed in the mid-1980s to help manage and assign IP addresses.

10.1.1   Reverse Address Resolution Protocol

RARP operates at the bottom of the network stack, at the hardware link level. A RARP client broadcasts a special packet with its network hardware address to some RARP server on the same subnet and receives back its assigned IP address. Most Unix systems come bundled with a RARP server daemon. Currently the RARP protocol is most commonly used by devices such as print servers that need little more than an IP address in order to function. Because it is implemented at the bottom of the network stack, RARP is generally not supported by micro TCP/IP vendors.

10.1.2   Bootstrap Protocol

BOOTP is an alternative to RARP and is used to configure networked micros. BOOTP, first proposed in 1989, uses the network application layer of TCP/IP and can deliver more configuration data in a single packet. BOOTP is commonly used by the TCP/IP stacks on Unix workstations, X-terminals, and desktop computers to manage and configure TCP/IP parameters such as names and IP addresses from a central location.

BOOTP uses the User Datagram Protocol (UDP) of IP and can be implemented as an application program. Relay agents in routers can forward packets across subnets to a distant BOOTP server. BOOTP extensions have been added and standardized since the protocol was first introduced, with the latest revision dated mid-1995.

BOOTP can provide a large amount of host-configuration data in a single packet exchange but requires an administrator to maintain a centralized database of BOOTP information for each client node based on the hardware address of the network card. When a new host is added to the network, the local hardware address of the installed network card is given to the person maintaining the BOOTP server(s). A unique IP address is assigned manually, and a new entry is added to the BOOTP data file with the associated configuration data.

10.1.3   Dynamic Host Configuration Protocol

BOOTP was designed for a stable environment in which nodes have a relatively permanent network connection—they change infrequently. With the increased use of portable computers, remote access, and wireless networking, it is possible for a computer to move to another location quickly— a situation not easily handled by BOOTP. In addition, a remote node connected through a serial port and modem does not have a unique hardware address.

The proliferation of TCP/IP-based networks, coupled with the growing demand for Internet addresses, has made devising a means of sharing and conserving a pool of TCP/IP addresses critical for corporations that wish to provide Internet access to their end users.

The Dynamic Host Configuration (DHC) working group of the Internet Engineering Task Force (IETF) was established in 1989 to develop techniques for configuring hosts dynamically. The Dynamic Host Configuration Protocol (DHCP) builds on the existing BOOTP protocol and supports all BOOTP vendor extensions. Where it differs from BOOTP is that the BOOTP server merely stores a preset configuration for a BOOTP client and delivers it on boot-up. DHCP, by contrast, automatically configures DHCP clients, using rules preestablished by the network administrator.

DHCP uses IETF's IP2 protocol, with its longer naming scheme, to support the growing number of TCP/IP nodes. BOOTP, a protocol used on many Unix systems for dynamically assigning IP addresses, does not support IP2.

Under DHCP, a computer is designated as the DHCP server. Computers on the network that do not have permanent IP addresses are DHCP clients. When the DHCP server is configured initially, it is given a block of IP address numbers that it can dispense to nodes that need IP addresses.

Figure 10-1
How a DHCP Server
Works

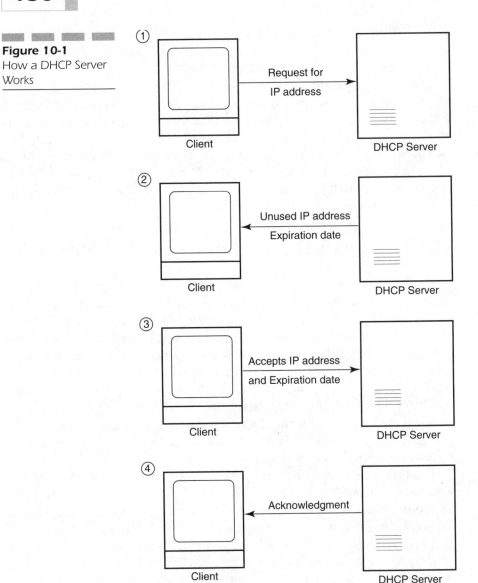

The process a DHCP server goes through to allocate addresses is illustrated in Fig. 10-1. When a new node comes onto the network, it broadcasts a request for an IP address. The DHCP server checks its table of address assignments, selects the next available address, and sends back a response to the requesting node. However, a requesting client must first find a DHCP server! Fortunately, the protocol is constructed so that a client may negotiate with more than one DHCP server.

A DHCP server could be put in each subnet, or there could be one central DHCP server to which all clients connect to get an IP configuration. The central DHCP approach requires that routers or some other relay agent be capable of forwarding DHCP packets.

The response sent back to the requesting node includes an offer of a specific IP address and an optional "lease period," which is the length of time the client may use the address. The client software decides which offered IP address to accept (often based on the lease time) and sends its acceptance to the offering server.

DHCP allocates addresses in one of three ways:

- *Automatic allocation.* The DHCP server assigns a permanent IP address (from a pool of IP addresses) to a DHCP client requesting an address.

- *Dynamic allocation.* The DHCP server assigns an IP address for a limited period of time (lease period) or until the DHCP client specifically relinquishes it, whichever comes first. A DHCP client can renew its lease before it expires and continue using the same IP address. Dynamic allocation allows a finite number of IP addresses to serve larger numbers of clients that are only intermittently connected. Leases are also useful for wireless networking, where the remote node will be crossing into a different cell and will need to reconfigure itself.

- *Manual allocation.* The IP address is chosen by the network administrator, but the DHCP server is used to convey the assignment to a DHCP client.

Automating the assignment of an initial IP address to the client makes it very easy to add a new client to a network. If a client moves from one subnet to another, DHCP can make the appropriate adjustments to the client's IP configuration. Dynamic allocation allows an organization to "time-share" a block of IP addresses among many clients, reducing the total number of IP addresses required.

With DHCP, nodes on a network are leased TCP/IP addresses as they log onto the system. Microsoft has built DHCP server capability into Windows NT Server (but not NT Workstation). Windows 95 has DHCP client capability. Apple's Open Transport includes DHCP client capability. Many of the third-party TCP/IP packages for Windows have DHCP client capability.

The IETF continues to define the DHCP standard, which will eventually give network managers a better method for conserving and sharing Internet addresses. The IETF will also address data storage and load balancing problems as it develops a new version of DHCP (version 6).

10.2   Access Security

Big companies should start with a firewall server stationed at the gateway point. Software solutions are fine for a small LAN. Tunneling software allows companies to use public Internet lines instead of leased lines to securely connect disparate internal networks.

Three common techniques are available to keep an intranet secure from unauthorized access from the Internet: IPX-to-IP gateways, firewalls and tunneling, and virtual private networks (VPNs). An organization's approach to intranet security should be based on the sensitivity of its data.

10.3   IPX-to-IP Gateways

The Internet and intranets use the TCP/IP network protocol for transportation. Therefore, the safest approach to connecting a network to the Internet is to avoid loading the TCP/IP protocol on any server that contains private information. An IPX-to-IP gateway is designed with that thought in mind.

The gateway runs as a service on a file server or on a dedicated PC. It converts all of the IPX requests to IP. Connections to the Internet can be handled without loading the IP protocol on each client PC; they only need to have IPX installed. IPX is not only for Novell's NetWare; Windows NT also bundles the protocol, and Novell's IntranetWare comes bundled with a gateway.

Unfortunately, although a gateway protects sensitive data and servers from intruders, it doesn't allow an organization to publish and protect its private data. Sharing information is one of the purposes of an intranet. An organization wants to make all its information available on the intranet but restrict access based on user needs. If users need to access data that is not on an intranet server running TCP/IP, they will have to connect to the network using the IPX protocol through a separate remote access. A gateway is a good solution only if the objective is to connect a network to the Internet.

10.4   Firewalls

A firewall is a network node that is set up as a boundary to prevent traffic from one segment from crossing over to another without authorization. Firewalls are used to improve network traffic as well as for security purposes.

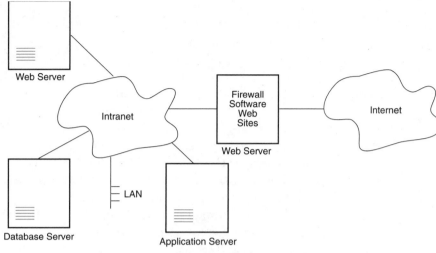

Figure 10-2
Firewall Implementations

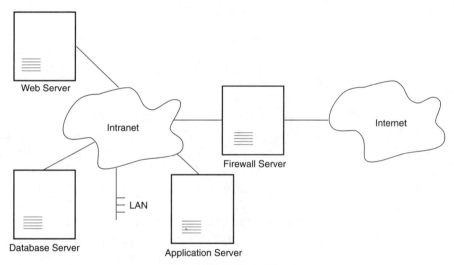

A firewall is a software program that sits on a Web server or runs on a dedicated computer, as illustrated in Fig. 10-2. The firewall software protects all of the intranet servers, including Web, file transfer protocol (ftp), and E-mail servers, on both sides of the connection. For example, an organization can prevent outsiders from accessing company data via the Internet, as well as control what internal users can access on the local network by routing all requests through a firewall.

A firewall sits between the internal network and the outside connection (usually a router) to the Internet. Today, more and more organizations are putting firewalls between LANs as well. A firewall between the finance department and the rest of the network puts a level of security on the data within that department's LAN, while still allowing the users in the department to access companywide data.

Firewalls can control access using various criteria, such as user name and password, IP address, Ethernet address, or even time of day. Most firewalls also offer an optional VPN package that can send encrypted information across the Internet. Most products include monitoring, logging, and notification features so that an administrator can see when someone is trying to gain access and who that user is.

Most firewall vendors have incorporated additional security technologies into their products and formed partnerships with other security vendors to offer complete Internet security solutions. These security solutions include encryption, authentication, antivirus protection, protection from misbehaved Java and ActiveX downloads, and even server load balancing.

Today's firewalls use one or more of three packet-screening methods:

- Packet filtering
- Application proxies
- Stateful inspection

Packet filtering.   Packet filtering examines the source and destination addresses and ports of incoming TCP and UDP packets and denies or allows packets entry based on a set of predefined rules. Packet filters are inexpensive, but configuring them is relatively complex. In addition, packet filters are susceptible to IP spoofing (see Sec. 11.6.3).

Application proxies.   Application gateways use application proxies, another type of firewall. These are programs written for specific Internet services, such as Hypertext Transfer Protocol (HTTP), ftp, and telnet, that run on a server with two network connections, acting as a server to the application client and as a client to the application server. Since they evaluate network packets for valid application-specific data, application proxies are generally more secure than packet filters. They also feature network address translation, which prevents internal IP addresses from appearing to users outside the internal network. However, performance suffers because of the double processing.

Stateful inspection.   A third type of firewall technology is called stateful inspection by Check Point Software Technologies. Like packet filtering, it first intercepts packets at the network layer, but then it inspects entire packets, comparing them against known bit patterns (states) of friendly packets. It generally results in slightly higher performance than application proxies.

One recently announced product is Secure WAN (S/WAN), codeveloped by RSA Data Security and TimeStep. S/WAN operates at the IP level and incorporates the Internet Architecture Board IP security standards. Its intent is to enable corporate customers to secure connections between their private networks and the Internet.

10.4.1 IP Security Protocol

IETF's IPSec (IP Security) protocol will enable firewalls from different vendors to establish a secure Internet channel—a virtual private network. IPSec includes facilities for encryption, authentication, and key management and can ensure privacy and authenticity over a public IP network. However, the overhead generated by the protocol's extension headers used to identify encrypted IP packets may cause a packet's size to exceed the size limit imposed by some routed networks.

IPSec, a part of IPv6 (see Sec. 5.2.2 for more information on IPv6), encrypts the tunnels so that an IPX packet running over a virtual private network, for instance, is secured. Virtual private network products that support IPSec will be able to communicate public keys and encryption algorithms with each other to set up virtual private network sessions.

10.4.2 Proxy Servers

Proxy servers are a type of firewall, a buffer between two networks. They allow companies to provide Web access to selected people by preventing unauthorized inbound traffic and restricting downloading by blocking specific Web sites.

Once a user's browser is configured to access the Web via a proxy server, the following procedure is followed each time the user chooses a page. First the browser accesses the proxy server and gives it the uniform resource locator (URL) of the page the user is asking for. The proxy server checks its cache of saved Web pages to see if it has a copy of the requested page. If not, it connects to the requested Web site, downloads the page, saves it to its local cache, and passes it on to the user.

When the next user asks for a Web page, the proxy server checks to see if it has a copy of the requested information; if it does, it sends that copy back to the browser. This eliminates the need to access the Internet, cutting

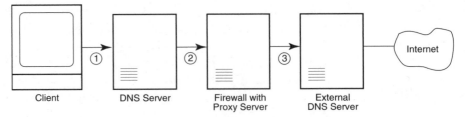

Figure 10-3
Proxy Server Handles
Conversion to Inter-
net IP Addresses

Client   DNS Server   Firewall with
Proxy Server   External
DNS Server

① Internal client asks internal DNS server to resolve Internet host name

② Internal DNS server passes request to firewall

③ Firewall's proxy server initiates DNS inquiry to external DNS server to resolve Internet host name

down traffic load and improving response. Because Web pages change so often, most proxy servers delete copies of pages older than a user-specified age or make a quick comparison of the local page and the Internet-based page. If the page on the Internet is newer, the proxy server retrieves the new page, caches it, and sends it to the user. If the page on the Internet isn't newer, the cached page is sent.

The proxy server can also log and report on what sites users are viewing. Because each request to access the Web passes through the proxy server, it is possible to examine each request and block inappropriate sites. The problem with this approach is keeping the list of "good" and "bad" sites up-to-date. Since each request is examined, it is possible to log all requests for later examination.

A proxy server is also used to convert internal station addresses to global Internet IP addresses, as illustrated in Fig. 10-3. Because users talk to the internal proxy server and not to the actual Internet, the network addresses on the internal network never make it to the outside world. All the Web traffic sent from users' machines to the Internet appears to have come from the proxy machine.

10.5   Tunneling

Encryption involves the encoding of data packets for their journey over a network. A firewall- or PC-based encryption package scrambles the packets using some sort of encryption algorithm so that they cannot be read if intercepted. The destination firewall or client machine decrypts the packets on the other end.

Tunneling, meanwhile, is encapsulation—one protocol encapsulates foreign data packets. Tunneling encapsulates packets of data written for one network protocol inside packets used by another. Tunneling is used to allow TCP/IP data to be transmitted over non-TCP/IP networks. Since IP networks, like virtual private networks, can handle only IP packets, other types of data, such as IPX, must be "tunneled" or packaged inside IP packets. This type of software can be used to create a secure intranet through the dial-up network, in effect creating a virtual private network. IP tunneling lets TCP/IP networks carry information packets for the other protocols it supports, and guarantees the sender's identity.

Tunneling technologies aren't new. Such two-key encryption systems have been around since the mid-1970s and have been used in client/server configurations. The server side encrypts something that can't be decrypted without the software on the client, and vice versa.

Many VARs use browser-based encryption schemes, such as Netscape's Secure Socket Layer (SSL). But these solutions are not nearly as comprehensive as tunneling software. Tunneling works at the IP level, so *all* traffic is encrypted—telnet, ftp, gopher, etc. Netscape's SSL does encryption at the application level, meaning that only Web information is encrypted.

Virtual private networks require secure or encrypted tunnels. There are two new types of tunneling protocols today that allow remote PPP (Point-to-Point Protocol) sessions to be accepted by an Internet service provider and authenticated by queries to corporate security databases: Point-to-Point Tunneling Protocol (PPTP), a technique promoted by Microsoft in its NT servers, and Layer 2 Forwarding (L2F), used by Cisco in its routers. These protocols allow Internet service providers and carriers to receive an incoming call from a remote user and tunnel it directly to a corporate LAN using the Internet.

10.5.1   Point-to-Point Tunneling Protocol

PPTP is a software solution. The Internet service provider does not have to support tunneling; the software running on the client or on the back-end server on the network does all the tunneling encapsulation. PPTP requires IP-based routing.

PPTP can be used to route PPP packets over any IP network, including the Internet. PPTP also tunnels or encapsulates multiple protocols, including IP, IPX, and NetBEUI, and can be used to send any type of packet over the Internet, such as NetWare IPX packets. Combine this technology with an IPX-to-IP gateway for a total intranet security solution. Employees will use a PPTP connection to encapsulate IPX and access the private data on internal servers, and customers will see only the data that is kept on an organization's public intranet servers running the TCP/IP protocol.

Unfortunately, the PPTP technology has a few drawbacks. First, the organization must be connected to an Internet service provider that offers PPTP. Second, it is a relatively new technology, and not everyone has agreed to use it. However, it is included with Windows NT, and vendors such as U.S. Robotics include it in their access servers.

10.5.2 Layer 2 Forwarding

L2F is hardware-based. It is not tied to a specific protocol and can run at a low network layer—Layer 2 of the OSI seven-layer model. L2F allows session parameters to be negotiated by dial-up clients with their own corporate gateway. Authorization, address negotiation, protocol access, accounting, and filtering are all controlled by the corporate network. Encryption runs transparently over L2F tunnels.

10.5.3 Layer 2 Tunneling Protocol

The trouble with PPTP and L2F is that they are proprietary. To support both protocols, organizations have to maintain two separate sets of hardware. A PPTP message can't be serviced by a L2F tunnel. This has hampered the acceptance of both protocols.

So Cisco and Microsoft actually put their rivalry aside and jointly developed a hybrid protocol called Layer 2 Tunneling Protocol (L2TP). L2TP allows the authentication and authorization process to be forwarded from the Internet service provider to a server located elsewhere on the Internet.

Like PPTP, the L2TP specification does not require built-in hardware support, although equipment that does support L2TP will provide faster performance and greater security than hardware that does not. Like L2F, L2TP is not bound to IP, so it will also work well over frame relay. Flow control was also added to L2TP to ensure that systems aren't fed more data than they can handle. The new specification also allows multiple simultaneous tunnels to be opened between endpoints. One thing still missing is standardized encryption, so interoperability may become problematic.

L2TP is in the draft phase at IETF and is expected to begin appearing in Cisco and Microsoft's products by early 1998.

10.6 Virtual Private Networks

A virtual private network is a network that uses the Internet to reach branch offices or trading partners and uses encryption technology to

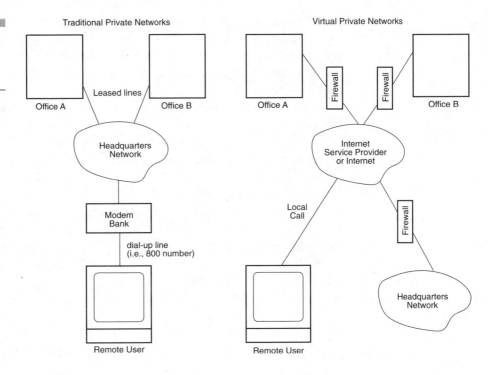

Figure 10-4
Traditional Private Networks vs. Virtual Private Networks

ensure safe arrival of the data. Another way to think of a VPN is that the Internet is used as a replacement or extension to private network services like frame relay and leased lines, as illustrated in Fig. 10-4. This generation of intranets looks and feels just like the Internet but is more secure and shoots packets through more quickly and efficiently. And it provides the infrastructure for an extranet, which incorporates business partners.

As illustrated in Fig. 10-5, VPNs are implemented through encryption

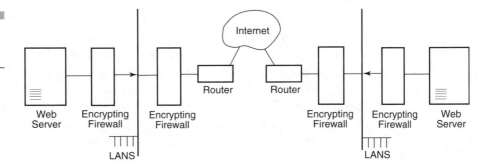

Figure 10-5
Virtual Private Network Architecture

and authentication features within firewalls and routers that are put together by network managers over their existing Internet connections. By using the Internet as their backbone, organizations get a flat rate rather than the unpredictable variable rates charged by telephone companies for frame relay.

The initial VPN products required the same brand of firewall on both ends. This requirement will be relaxed with the acceptance of the new IPSec VPN standard (see Sec. 10.4.1). VPN providers will be able to add companies and partners easily as long as their firewalls or routers support IPSec. However, because of the newness of the standard, little interoperability testing has been done.

Carriers and Internet service providers see this as the next logical market for their private IP and frame relay services. But they have their own ideas about what a VPN should be. Bell Atlantic, for instance, considers a VPN to be everything from a leased-line fiber distributed data interface (FDDI) to frame relay and an asynchronous transfer mode (ATM) network. AT&T calls its IP- and NetWare-based WorldNet Intranet Connect Service (WICS) a VPN. Neither provider includes encryption in its services, which is what really sets a VPN apart from an intranet. Tunneling is another feature still on the horizon for these services.

10.6.1   VPN Offerings

AT&T   AT&T's WICS is an IP backbone that supports both IP and IPX traffic natively, using Web or NetWare-based servers. AT&T offers dial-up, leased-line, or frame relay access to WICS, which is considered a subset of the company's WorldNet public Internet network. The security piece is based on Novell Directory Services (NDS), which authorizes users and their access to network resources. Although this doesn't constitute security in pure VPN terms, it does offer more specific and detailed security than traditional Internet transmission methods. NDS lets a network administrator set up security provisions for specific files on a server, which is a level of security not found on a general IP network.

MCI   MCI is bundling its WebSite, Managed Services, HyperStream, Internet, and SafeNet encryption services into a comprehensive VPN offering. These services can be combined for intranet, VPN, and extranet networks. MCI's WebSite service can be used for an extranet, for example, either over the Internet with turnkey browsers or using MCI's own frame relay or ATM services for higher-performance connections. MCI's Managed Services is basically an intranet service.

Sprint   Sprint considers its Dial IP intranet service the foundation for a VPN offering. Dial IP can be packaged with Sprint's Managed Firewall Service, which is composed of Raptor Systems Inc.'s firewalls and client software that encrypts traffic as it goes across the Sprint IP intranet network.

ANS   ANS's Virtual Private Data Network (VPDN) comes with end-to-end encrypted routing using the company's home-grown filtering routers as firewalls. Unlike some services, ANS integrates and manages most of the elements. ANS, like most service providers, plans to offer Reservation Protocol (RSVP)- or ATM-based quality of service options in the future. VPDN can be used to build an extranet as well. The ISP also offers an intranet service without the encryption, and InterLock, its managed firewall service.

PSINet   PSINet Inc. sells its Intranet Service as a turnkey package, complete with routers that plug into PSINet's network for connecting branch offices to the corporate intranet. The Intranet Service runs on a different set of routers from those used for PSINet's public Internet service. The backbone is based on frame relay, but the service is all IP, not frame relay. The frame relay switches make private paths through the network to link all the customer's sites together. All a customer's packets moving back and forth touch only the routers on that customer's premises.

There are no firewalls with the service—PSINet considers its packet-filtering GT Secure routers, from Proteon Inc., as well as the network configuration, security enough. Security is designed into the service by logically isolating customer traffic. PSINet feels that firewalls can affect network performance and inadvertently make a single point of failure, with branch offices going through one firewall point.

UUNet   UUNet's extranet services, ExtraLink and ExtraLink Remote, were released in mid-1997. Two of the more innovative features of the UUNet extranet family of services are its guarantees of 99.9 percent availability and 150-ms end-to-end latency. If those levels are not met during a particular month, UUNet offers a 25 percent refund of the total price of the service for that period. All traffic is encrypted at the customer's site, which UUNet manages and monitors. UUNet's ExtraLink Remote, its dial-up extranet service for remote users, comes with client encryption software as well as token authentication to the corporate firewall. Like most service providers, UUNet stays away from implementing its customers' security policies and actual authentication parameters. When a company wants to add a business partner to the extranet, it merely contacts UUNet, which then gives that partner a contract for the portion of the service it uses.

10.6.2   Where Are VPNs Headed?

VPNs are the wave of the future for carriers and Internet service providers. For carriers, it's finally a chance to dominate Internet and IP services with a brand of service that taps into the carriers' strong suits: performance and reliability—not characteristics usually attributed to the Internet. Carriers are counting on their household names in this market. They are pushing their ability to deliver all the different pipes and services together—dial access, X.25, electronic data interchange (EDI), X.400, and Internet. One link will do it all.

Internet service providers will draw heavily from their IP heritage to win over customers and are now positioning themselves as extranet and virtual private network solutions providers rather than plain-old-vanilla Internet access companies. Their strategy is to provide an integrated solution that includes data, voice, and video. They aren't interested in selling pipes or boxes.

There is a big catch to these intranet, virtual private network, and extranet services, however. Security and performance are guaranteed only within a particular service provider's network. Once a user hops off one network and onto another—either the public Internet or another private IP service—traffic is vulnerable, and there's no telling how fast it will get to its destination.

For network managers, it's all about a cheaper and more comprehensive alternative for linking branch offices, traveling or at-home users, and, in some cases, business partners. A link through the Internet or a virtual private network through a service provider can cost as little as one-tenth as much as dedicated leased lines.

To date, there are few, if any, true VPNs up and running. The closest thing to a VPN in most cases is a frame relay network running IP traffic or an intranet or extranet. Most users pioneering in VPNs have built their own out of VPN equipment and public Internet services, for instance, mainly to give far-flung branches and on-the-road workers a way to reach the corporate network over the public Internet.

And if a VPN is defined as an encrypted service over the Internet, then true VPN services are not available today from carriers. Most offer intranet services with IP traffic running over frame relay, for instance, and no real encryption. Many of these early IP services have no quality of service features just yet—most are awaiting software that recognizes RSVP (Resource Reservation Protocol), which is not expected until sometime in 1998. In addition, encryption and tunneling techniques aren't yet standardized, making multivendor interoperability a problem. Also, federal law prohibits U.S. companies from using strong encryption on international links.

One big plus for the carriers' VPN-to-be services is that they offer more predictable performance than public Net services. But the bottom line is that it's still IP. Because of IP's connectionless nature, it is not as predictable as connection-oriented technologies, such as ATM. The only hope for quality of service in a VPN is RSVP and frame relay Committed Information Rates (CIRs). RSVP will allow minimum bandwidth requirements to be set across the network. RSVP and CIRs are discussed in Sec. 14.5.1.

Even with all the proposed improvements to VPNs, some ISPs still don't recommend running mission-critical applications such as transaction processing, for example, over the Internet for performance reasons. They maintain that the infrastructure of the Internet itself is still not reliable enough for time-sensitive, mission-critical applications. They advise organizations to wait for more security and better performance before they take that plunge.

10.7 Web Access

The simplest form of Web access is to use HTML pages that statically reference text, images, and audio. This approach is easily implemented but inflexible—the information on the page doesn't change. The next evolution includes page-embedded scripts in the HTML pages so that pages can be formatted differently, depending on input parameters. Both of these methods require minimal expertise. However, this "document-centric" approach does not lend itself to displaying data.

With the common gateway interface (CGI), pages can be dynamically built depending on data contents. CGI can dynamically format pages using application logic written in PERL, C, or C++. As the number of requested objects grows, directory maintenance becomes more cumbersome, and organizations turn to databases to manage the Web-site data.

For lower request volumes, a single server is usually satisfactory. As the workload increases, more processors and memory can be added. However, as the workload increases, performance declines because a new process must be spawned for each request. Newer Web applications interface with middleware, which uses threads to minimize overhead and provides a way to retain context-state data for complex transactions.

Transaction-processing monitors are seeing a major comeback to handle load balancing and transactions that span distributed servers. Because the HTTP protocol is stateless and servers can't track the client execution context, TP monitors are used to maintain the current state of each transaction as it progresses through different parts of the application logic.

TP monitors can also route requests around failed nodes and prioritize messages. TP monitors can funnel multiple client requests into fewer sessions, thereby reducing database management system memory requirements. The two-phase commit protocols of TP monitors can support transactions across several nodes running different DBMS products, ensuring transaction integrity across heterogeneous databases.

10.7.1 Common Gateway Interface

In the traditional approach to i-net applications, a very thin client communicates to a Web server via HTTP. The server's main purpose is to serve documents or files that can contain very sophisticated graphics that are files in themselves. The presentation of these documents occurs via a thin, portable, and often free client over a TCP/IP network connection. For its intended purpose—page description—HTML is fine. But for customized features, HTML is not much help. These features are implemented on the server as a common gateway interface.

CGI is the standard that emerged to support data access. CGI has a client portion and a server portion, as illustrated in Fig. 10-6. A CGI script is used by Web server software to communicate with other applications on the server or on the network. It is used to connect to databases and other back-end services, for moving text from one server or platform to another. It is this software, a CGI script, that gives Web servers capabilities such as forms processing, image mapping, and links to database software and other packages. CGI scripts are executed by the server in response to user actions.

Figure 10-6
Common Gateway
Interface Structure

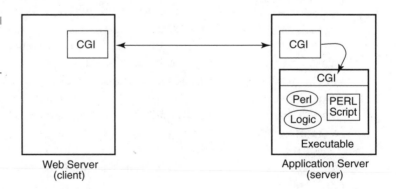

Web Server
(client)

Application Server
(server)

CGI is not specific to database access; rather, it's a generic tool for interfacing a Web browser with an executable program running on a Web server. CGI programs receive parameters and return results by executing the program. The results must be formatted as HTML for the browser to display them. This data-access architecture was the first available and currently dominates the market.

For example, a Web page might contain a searchable product catalog. A user can enter a keyword into an HTML form to look for a product of interest. The CGI application sends the request to a database program for execution and tells the database software to return the results of the search as an HTML page. This CGI application generates a new page each time the database is searched. CGI applications are also used by Webmasters (the administrators responsible for the management and often the design of Web sites) to count Web-page "hits" (a hit occurs when a user stops at a site) and send the data to a spreadsheet program for analysis and reporting.

CGI actually refers to the predefined way in which these programs communicate with the Web server, but it has come to refer to the programs themselves. The original CGI scripts were written in PERL or some other script or high-level language. Java, JavaScript, and VBScript are primarily used now.

As applications require more functionality and higher graphic and response interfaces for the desktop, the thin client becomes fatter. HTML, although capable of presenting highly graphic materials, does not have the interactive capabilities, such as pull-down menus and list boxes, that today's business users expect. In striving to be portable, HTML sacrificed some of the interactive capability that is proprietary to each operating system.

Plug-ins fill in these gaps. Plug-in technologies execute on the desktop while the user is in the browser and enable much more interactivity than HTML. In what is known as the "download once, run many times" method, the executable code for the plug-in resides on the server and downloads to the client when necessary. It then remains on the client's hard drive waiting to be called into action by its browser.

However, plugs-in do have drawbacks. For example, they must be downloaded "just-in-time" when needed. And both browsers and clients suffer from their proprietary nature, so plug-ins will probably be replaced by Java and ActiveX.

Regardless of how CGI scripts are generated, the Web development environment using these tools lacks the ability to maintain "state" between actions, a precondition of database access and transactions. HTML and CGI are built to execute independent actions; HTML invokes individual tags or commands, and CGI executes a program and expects a result. But in database access, the database opens, prepares an operation, executes it, processes the resultant set, and closes in sequence.

When a typical database connection is initiated, the query state is maintained during processing. Every time CGI executes code, it must manage all phases of database access. It can't simply open the database, ready a query, and execute it the next time the program is run. This constant "opening" and "closing"—as well as the preparation and execution of the query—creates a lot of overhead.

In addition, each user's invocation of a server-resident CGI script creates a separate process on the server. CGI technology runs out of steam when the number of users increases because each process requires its own resources. Consequently, CGI applications compete with the Web server software for resources, such as CPU time, disk access, and network bandwidth.

CGI applications are potential security risks because they can do things that servers don't normally allow users to do. A well-designed server will provide a way for CGI applications to comply with its internal security measures before it returns a file to a user, but it is easy to code a CGI application to ignore these security measures and access whatever file it wants.

In addition, database management software expects users to stay connected. Web pages do not stay connected to remote applications.

Most developers do not consider CGI a viable solution for high-volume, transaction-oriented systems. Organizations that are attempting to develop intranet implementations for transaction-oriented applications are looking to other software (currently groupware) for the necessary support.

10.7.2 NSAPI, ISAPI, and ICAPI

An alternative to extending the abilities of the server is to use the server's application programming interface (API). The API allows the developer to modify the server's default behavior, which forces the server to use the code in an API module instead of its own built-in code. Code is encapsulated in a dynamic link library (DLL) and linked dynamically at run time by the server. When functionality is added to a server, the API acts as an interface between the server and another application.

CGI is slow. Java and JavaScript are also slow. Because APIs are lower-level interfaces, they will run faster. APIs can be used to connect directly to a Web server, process HTML streams inside the Web server, and modify server behavior.

APIs can be used to share data and communications resources with a server, share function libraries, and provide capabilities for authentication and error handling. Another advantage of APIs is their ability to maintain state. Because an API application remains in memory between client requests, information about a client can be retained and used again when the client makes another request.

However, unlike the case with CGI, API functions are server-specific because each server has a different API. Developing the code is more complex, as the API must manage multiple process threads and clean up memory after it is run.

Web server vendors have also created APIs for server-resident applications. Netscape Server API (NSAPI), Microsoft Internet Services API (ISAPI), and O'Reilly's WebSite API were all designed to improve performance by eliminating the process creation overhead of CGI. They do improve performance, but the application has to fit into the server's address space—an expensive proposition. In addition, these APIs are proprietary and do not specifically address database access.

NSAPI NSAPI allows a browser to execute programs on a Netscape Web Server. Using NSAPI function calls, Web pages can invoke applications on the server, typically to access data in a database. NSAPI can be used to modify almost every aspect of server operation, but its use is not a trivial task. Familiarity with how the server responds to requests is a prerequisite. Knowledge of the data structures and public functions that will be interfacing with the server is critical.

NSAPI DLLs can be written in any language that allows the creation of DLLs, such as C, C++, and Delphi.

ISAPI ISAPI from Microsoft and Process Software allows browsers to access remote server applications set up on Microsoft Internet Information Servers (IIS), as illustrated in Fig. 10-7. The executable programs are set

Figure 10-7
ISAPI Architecture

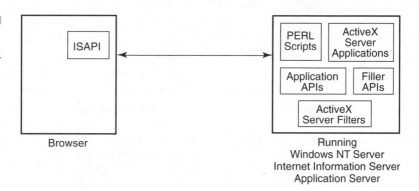

Browser

Running
Windows NT Server
Internet Information Server
Application Server

up as DLLs. They are loaded once and can serve multiple users. Using ISAPI calls, Web pages can invoke these DLLs on the Web server. IIS also includes an ISAPI DLL that allows queries to be embedded in Web pages that can access Open Database Connectivity (ODBC)-compliant databases on the server.

ISAPI interacts with the server through an extension control block, a data structure that contains information about how the API processes a request. The ISAPI consists of filters, which process, set up, or change requests before a server receives them, and applications.

ICAPI IBM's Internet Connection API (ICAPI) provides control over IBM's Internet Connection Server product. It is basically a clone of NSAPI for IBM's server product.

10.8 Security

The Internet is made up of private lines, and securing transmitted data over those lines is a major concern for organizations doing business over the Internet. Encryption is used to provide security when transmitting sensitive documents, such as legal contracts, medical records, and other confidential communications, over the public network. Without the use of encryption technology, E-mail communications can be intercepted and read using widely available packet-sniffing software. Encryption technology "scrambles" E-mail messages and ensures that messages can be read only by the intended recipient.

Data encryption, the most common transmission security solution, can be symmetric or asymmetric. Symmetric encryption is based on the Data Encryption Standard (DES), which uses the same 56-bit key on both ends of the transmission. Asymmetric encryption has been used for about a decade and uses public/private key pairs. The most prevalent encryption method in use today is the RSA algorithm from RSA Data Security. Current key lengths range from 48 to 1024 bits, but the longer the key, the higher the computational overhead for encryption.

The Secure Socket Layer protocol pioneered by Netscape combines encryption and authentication with message integrity checking. SSL supports secure and nonsecure messages in alternate transmissions, reducing overhead when sending the nonprivate portions of the transaction. Secure HTTP also provides application-level encryption and offers a choice of algorithms (DES or RSA) with different digital signatures.

User authentication is another major security issue. As security becomes

Figure 10-8
Security Layers for
Authentication

Security Layers	Examples
Something you *know*	Passwords
Something you *have*	Magnetic cards or tokens that are read electronically by the client machine
Something you *are*	A biometric physical characteristic or a third-party certificate authority such as Verisign

more and more of an issue, organizations are moving up the security layers for authentication, as shown in Fig. 10-8.

Intranets may seem more secure because they are behind a firewall, isolated from the Internet, and self-contained within the organization. However, intranets usually provide access to sensitive company data. Physical security is not as much an issue as is access security. Organizations should still use SSL for secure transmissions within the intranet and have the ability to restrict access to the file or field level. Most Web servers provide SSL support. Lotus's Domino provides field-level security restrictions.

Firewalls are usually thought of the way to isolate an intranet from the Internet. They are also used to isolate intranets from each other or remote access sites from the rest of the intranet.

Security issues are discussed in more detail in Chap. 11. Firewalls are discussed in Sec. 10.4.

The tradeoff for high levels of security is processing time and infrastructure costs. Each level of security that needs to be checked adds processing overhead to the transmission—at both ends. This becomes an especially critical issue for electronic commerce because the cost of authenticating an on-line transaction is currently too high. Consequently, much work is being done to develop a secure electronic cash technology that can handle such transactions cost-effectively. Security as it relates to electronic commerce is discussed in Chap. 17, "Electronic Storefronts," and Chap. 19, "Web-based Electronic Data Interchange."

11

I-Net Security

I-nets must be designed with the idea of safeguarding the information resources of the client and the server. I-net security is crucial, even if data is not sensitive. Not only can intruders look at the data, they can delete it or change it if they want.

Any outsider can infiltrate internal company systems and network devices via an Internet connection. Once inside, an intruder can find ways to look around, change or steal data, destroy data, or just cause general havoc. Even E-mail is not secure. Anyone with a protocol analyzer and access to routers and other network devices that handle E-mail as it travels across the Internet can read and even change messages unless they were sent securely.

Internet connections will never be 100 percent secure. Organizations need to assess the value of the information they are trying to protect and balance that against the likelihood of a security violation and the cost of implementing various security measures. In addition, the more locks there are, the more time it takes to let things through. A possible security architecture is illustrated in Fig. 11-1.

Figure 11-1
Security Architecture

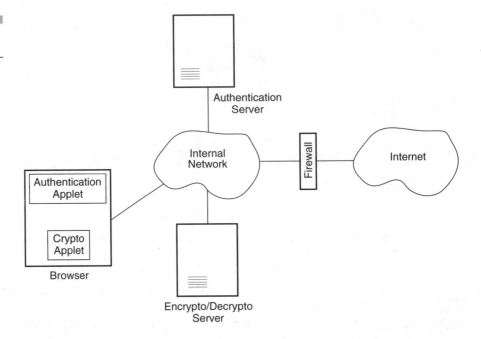

Because TCP/IP was not designed with security in mind, additional technology and policies must be used to solve typical security problems such as authentication, privacy, tamper-free messages, and protection.

A single technology already exists that provides the foundation for solving all of these challenges: cryptography. Cryptographic technology is embodied in industry-standard protocols such as SSL (Secure Socket Layer), SET (Secure Electronic Transactions), and S/MIME (Secure Multipart Internet Mail Encoding). These standards provide the foundation for a wide variety of security services, including encryption, message integrity verification, authentication, and digital signatures.

How digital identification works within a network is illustrated in Fig. 11-2. A document is encrypted with a public key. A message digest is pro-

Figure 11-2
How Digital Identification Works

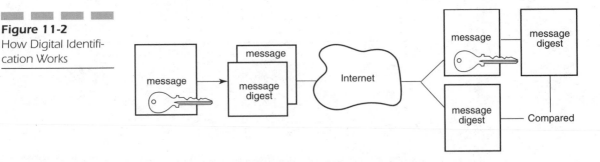

duced, which is a condensed text string that has been derived from the contents of a text message using a one-way hash function. The message and the digest (also encrypted) are sent over the Internet. When the message is received, a message digest is again created. The message is decrypted using the sender's public key, and the two message digests are compared.

11.1 Cryptography

Cryptography comprises a family of technologies that include the following:

- *Encryption* transforms data into some unreadable form to ensure privacy. Internet communication is like sending postcards—anyone who is interested can read the message. Encryption offers the digital equivalent of a sealed envelope.

- *Decryption* is the reverse of encryption; it transforms encrypted data back into the original, intelligible form.

- *Authentication* identifies an entity, such as an individual, a machine on the network, or an organization.

- *Digital signatures* bind a document to the possessor of a particular key and are the digital equivalent of paper-based signatures. Signature verification is the inverse of a digital signature; it verifies that a particular signature is valid.

All of these technologies make use of sophisticated mathematical techniques.

11.2 Message Security

Transmissions within an intranet need to be just as secure as those that are sent outside. Three fundamentals of Web-server security are fortification, authentication, and encryption. Firewalls provide the fortification. Security protocols can handle the authentication of strangers and encryption of messages. Digital signatures are used for authentication.

A security protocol is a communication protocol that encrypts and decrypts a message for on-line transmission. Security protocols work on different network layers and use different encryption algorithms, but share several concepts. The client and server communicate in such a manner that someone watching the communication on the network is unable to decipher what the client and server are saying to each other. Authentication, using certificates, makes sure the data hasn't been altered or forged.

Digital signatures are electronic signatures that cannot be forged. This coded message is transmitted with the text message. It ensures that the document was sent by the "person" signing it and has not been altered since it was signed.

Firewalls do their own authentication, using IP addresses which are assigned to each server, client, and network device, and can be spoofed. Passwords are the most common method of authentication used today, but users tend to make poor password choices that can be guessed by an experienced hacker. In addition to passwords—"something you know"— organizations are turning to "something you have," such as tokens or smart cards.

Tokens are small credit-card-sized devices that remote users carry around. When a user attempts to connect, an authentication server on the network issues a challenge, which the user keys into the token device. The device displays the appropriate response, which the remote user then sends to the server. Smart cards are similar to tokens but require a smart card reader to process the challenge.

11.2.1 Encryption

To send secure messages to a server, the sender needs to be able to use the encryption services of that server. Encryption is the process of scrambling digital information so that it can be decrypted by anyone holding a password, or "key." In private-key encryption, data senders and receivers use the same key to encrypt and decrypt messages. In public-key encryption, each party holds a common public key and a private key known only to the individual user. The two keys are related by a mathematical algorithm which lets each side of the pair identify a message from the other.

The public key is available to anyone, easily downloaded from the organization's Web page. The Webmaster has the private key. Any public key can encrypt messages, but only the holder of the private key can decrypt and read them. The public key can also decrypt messages, but only if they were encrypted by the private key.

Key escrow, encrypting keys with a corporate key, is usually used by organizations to provide an additional level of accessibility. If an employee's private key becomes unavailable for whatever reason, the key can be decrypted with the corporate key. If a person leaves or is fired from a company, the company can then decrypt the employee's key and use it to gain access to company files.

Figure 11-3
Public-Key Encryption

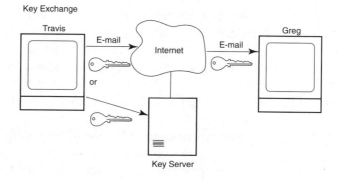

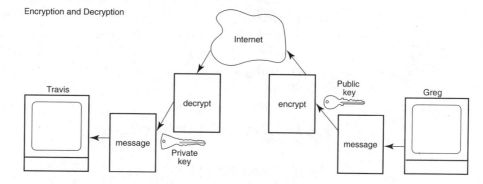

11.2.2 Public-Key Encryption

Public-key encryption, a recent improvement over private-key encryption, lets individual private-key holders send encrypted messages with digital signatures, which can be validated using a public key available to anyone and then later decrypted using the recipient's private key. The process is illustrated in Fig. 11-3. Travis sends his public key to Greg or places it on a key server accessible via the Internet so that Greg can retrieve it. To send a message, Greg encrypts his E-mail message using Travis's public key. Travis decrypts the E-mail using his secret key.

The leading public-key algorithm in use today is RSA from RSA Data Security Inc., which uses two separate keys. Each participant in an encrypted transaction has one private key, known only to that person, and one public key, which can be seen by anyone. The same key cannot be used for both encryption and decryption. There are fewer keys to manage, but RSA requires lots of processing power.

The other key algorithm used is the Data Encryption Standard (DES). DES is a 20-year-old, 56-bit coding scheme that is used by many financial institutions. Government agencies use a mandatory security solution known as Fortezza which uses a minimum 56-bit key based on DES. With Fortezza, each Internet user is assigned a unique identification string which is stored on a credit-card-sized microprocessor called a token. These tokens, which serve as personal Internet keys, are carried by each federal employee. To use Fortezza, all Internet workstations and Web servers must be equipped with devices that can read these tokens. To use the tokens, users must enter a password.

In the summer of 1997, an ad hoc team used thousands of computers to crack a message coded with DES in response to a challenge from RSA Data Security. It took four months and 14,000 computers to try out every possible combination for the keys. While hackers are unlikely to have such resources at their disposal, this does reinforce the fact that if someone wants to break in badly enough, he or she will find a way.

In response to the news, most institutions using DES announced that they were using a stronger version of DES than the one that was used in the challenge.

11.2.3   Digital Signatures

A digital signature is an electronic signature that cannot be forged. It's the network equivalent of signing a message—you can't deny you sent it, and the receiver knows it came from you. It is a coded message that accompanies a text message transmitted over a network. Digital signatures are usually implemented using the de facto standard RSA public-key cryptography method.

The digital signature's use with messages is illustrated in Fig. 11-4. To send a digital signature, the sender's software uses an algorithm to compute a hash value from the text message. Using the sender's private key, the sender's software encrypts the hash value, turning it into a message digest.

Figure 11-4
Digital Signature
Process

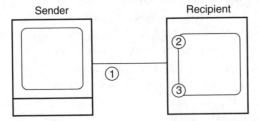

① Signs with private key and encrypts with sender's public key

② Verify signature with sender's public key

③ Examine for tampering and decrypt with recipient's private key

The text message is then encrypted with the private key, and both message and message digest are transmitted to the recipient. The recipient's software uses the sender's public key to decrypt the message and message digest. Using the same hashing algorithm, a new message digest is computed from the text message. If it matches the one sent with the text message, the signature is authenticated.

A digital signature ensures that the document originated from the person signing it, and that the document was not tampered with after the signature was applied. However, the sender could be an impostor. To verify that the message was indeed sent by the person claiming to have sent it requires a digital certificate, which is issued by a certification authority.

11.2.4 Digital Certificates

Digital certificates verify to both parties in a transaction that the holder of a public or private key is the person or organization it claims to be. These electronic credentials are based on security standards, protocols, and cyptography techniques that establish an individual's or server's identity. Digital certificates bind owners to a pair of electronic keys that can be used to encrypt and sign information, as illustrated in Fig. 11-5.

The Internet Engineering Task Force's (IETF's) X.509 is a standard specification for a certificate which binds an entity's distinguished name to its public key through the use of a digital signature. The specification also contains the distinguished name of the certificate holder.

The key user's identity is attested to by a certification authority. Currently there is only one certificate authority open to the public, VeriSign

Figure 11-5
Digital Identification

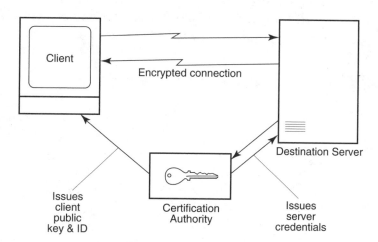

(www.verisign.com). In a little more than two years, VeriSign has issued over 500,000 digital certificates.

VeriSign has three classes of certificates. Class 1 verifies that E-mail bearing the user's address actually came from that address. For a Class 2 Digital ID, VeriSign checks a user's identity information against a commercial credit database. For a Class 3 Digital ID, the user must take identification documents to a notary, either a public notary or a VeriSign-approved local registration authority.

VeriSign recently announced its NetSure Protection Plan, which protects VeriSign Digital ID users against losses due to theft, impersonation, or corruption of their IDs, or losses from being unable to use their ID. Protection levels vary for each class of ID: Class 1 consumer IDs, $1,000; Class 2 consumer IDs, $25,000; Class 3 consumer IDs, $50,000; and Class 3 organization IDs, $100,000.

Other recent enhancements to the product include partnerships with popular Web directories that will let users list their Digital IDs, secure E-mail, and a universal log-in feature that will work at Web sites that accept VeriSign's Digital IDs. With the universal log-in feature, Digital ID users can use their IDs instead of user names and passwords to log in to all sites that accept VeriSign IDs.

On the horizon is Cylink (www.cylink.com), which is developing certification-authority services for the U.S. Postal Service that include a public certification authority. In addition, GTE, with its CyberTrust product line (www.cybertrust.gte.com), and BBN, with its SafeKeeper Certificate Management System, are supplying systems and software to businesses that want to establish their own private or in-house certification authorities.

The use of digital certificates took a giant leap in the fall of 1997 when Netscape shipped its Communicator 4.0 Web client with support for X.509, Version 3, a digital-certificate standard. Microsoft also built X.509 support into Internet Explorer 4.0, released in the fall of 1997.

11.3   Smart Cards

A smart card is a credit-card-like device with an embedded microchip. The card slips into a smart-card reader that is either standalone or installed in a keyboard. A sample architecture is illustrated in Fig. 11-6. The administrative server runs the public-key server software as well as the smart-card management system. The directory server is the repository for public-key certificates.

The chip is used to store information to support applications such as network security and electronic shopping. Several companies are develop-

Figure 11-6
Smart-Card Architecture

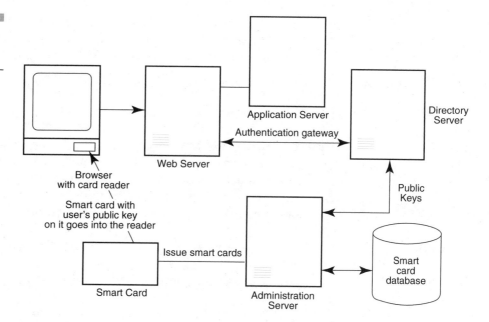

ing "smart-card" applications that would hold not only a user's profile and network-access privileges, but also financial and health-care information, and could also be used for shopping over the Internet. They provide authentication services for electronic commerce, replacing user identification and password combinations, and providing an alternative method of paying for electronic purchases (see Sec. 17.3).

Smart cards, nearly two decades old, have been used mostly in Europe and Asia, where they are used as stored-value cards in place of credit cards. The cards hold financial information as well as personal information. Smart cards in use today include telephone calling cards, other types of debit cards, and personal health records. Smart cards may replace electronic wallets (the term applied to software that provides browser users with digital certificates and payment systems).

Several pilot tests of the use of the smart card for Web security are under way in U.S. corporate environments. These cards usually use a special processor that can be customized to perform special operations, such as encryption. The capability to process differentiates these newer cards from their memory-only cousins.

Devices are now available from companies such as HP, Verifone, Netscape, and Microsoft that will put a public-key certificate on a credit-card-sized plastic card containing an embedded silicon chip. Such a device could simplify Internet security, including authentication between parties;

access control to Web sites and virtual private networks; and privacy protection for E-mail, transported files, and data stores.

Encryption technology scrambles data, allowing it to be accessed only by the intended recipient. Keys stored on a smart card are said to be secure because smart cards have built-in security features and are tamper-resistant.

Organizations are also looking at smart cards as a way to manage digital certificates. Netscape adopted a hardware standard for including digital certificates in smart cards, PC cards, and other hardware tokens. Microsoft is building digital certificate support into Internet Information Server 4.0—users will be able to issue and manage their digital certificates. However, digital certificates are vulnerable to viruses and to tampering with the hard drive, and they cannot easily be transferred among the many computers that some workers now use at the office, at home, and on the road.

A crypto smart card, which costs less than $2 to make, can be taken anywhere, but each computer needs a reader that a smart card can be inserted into or swiped through. Keyboards equipped with readers began to hit the market by the end of 1997. Some companies, such as Fischer International Systems Corp., are developing devices that allow smart cards to be inserted into floppy drives. Other companies, such as VeriFone and Litronics, are providing boxlike smart-card readers.

One of the major benefits of smart cards is portability. A smart card can be used to identify people to a corporate network regardless of where they are located (including on the road!) and what computer they are using. This eliminates the need to transfer a cryptographic key from computer to computer. By storing the private key on a smart card, a user can also create a digital signature for signing E-mail regardless of location.

Smart cards are emerging as a key way for IT and network managers to authenticate remote users. We may all be able to walk up to a kiosk in an airport, hotel, or some other public place, access hardware there using our smart card, and get our E-mail—all without even taking our laptops out of the case and finding a phone jack.

Smart card development has been hampered by the lack of standards. Card sizes, placement of contacts for reading card contents, and interactions between cards and card readers have been standardized in the ISO 7816 smart-card protocols. The EMV specification developed by Europay, MasterCard, and Visa has become a de facto standard for electronic payment systems, ensuring that all smart cards operate across all card terminals and related devices regardless of location, financial institution, or manufacturer.

Standards for operating systems for card microprocessors and application programming interfaces (APIs) to use the cards are not standardized yet. Sun Microsystems recently released Java Card API, which was endorsed by

Citibank, First Union National Bank, Schlumberger, and other manufacturers of smart cards, as well as by VeriFone and Visa. Java Card 2.0 allows a single terminal to accept cards issued for different purposes, and cards of different brands will be able to share a common infrastructure.

The PC/SC (PC and Smart Card) Workgroup, headed by Bull CP8, HP, Microsoft, Schlumberger, and Siemens Nixdorf Information Systems AG, has released its own specifications for smart-card support on personal computers. The specifications ensure interoperability among smart cards and smart-card readers and provide high-level APIs for application developers.

A consortium of industry leaders that includes IBM, Netscape, Sun, and Oracle's Network Computer division recently announced OpenCard, a new framework that sets interoperability standards for smart cards, readers, and applications. The framework incorporates PKCS-11, a public-key industry standard for cryptographic tokens. OpenCard does support the Java Card API.

Analysts feel that the rush to crypto smart cards will be slowed by a lack of agreement on standards. Nonetheless, financial institutions that are already doing business on the Web are expected to be among the early adopters of this technology. Visa has already announced that it plans to replace the magnetic strips on its cards with Java-based computer chips beginning in 1998. The chip on the card will include a processor and a Java Virtual Machine to interpret the Java code for each different machine it is used in.

Visa expects to have 2 to 3 million Java-based smart cards in use in the United States by the end of 1998 and 200 million worldwide by 2001. Consumers will still be able to buy things on credit and receive cash advances. In addition, they will be able to use their cards as electronic cash at specially equipped highway toll booths and vending machines after transferring cash to the card from an automated teller machine. They will also be able to use the cards as identification and authentication tools. If Visa wants to add features to its card, users would take the Java card to an automated teller machine and have these features downloaded.

MasterCard is working to establish a new electronic cash system which should be compatible with Java. American Express and Novus Services's Discover Card are also looking to convert to smart cards but haven't publicly announced the details yet.

11.3.1 Smart-Card Examples

Schlumberger Electronic Transactions Inc. offers a security application named SafePak that includes a smart card, a reader, and server software.

With SafePak, administrators can set up a system that accepts cash or data from a client based on the authentication in the smart card, which authorizes and encrypts transmissions.

Schlumberger has also teamed up with Pretty Good Privacy Inc. to market the Cryptoflex smart card, which includes 128-bit encryption and directory services.

Visa is working on a multiapplication smart card with Gemplus Corp. Multiapplication cards hold more than one function, such as cash purses and health-care information, on a single smart card.

VeriFone offers Personal ATM for digital cash. The user slips the card into the device, makes a connection with the bank, and keys in the amount to be downloaded. The card is then used to make purchases on any payment terminal that will accept it (like ATM machines) or over the Internet with a reader connected to a computer.

11.4   Security Protocols

A protocol is a set of rules. Security protocols detail how messages are "wrapped" to make them secure. What follows is a discussion of the major security protocols currently in use. However, none of them is bulletproof. When it comes to security, the sender needs to feel that something is being done to make the transmission—be it a credit-card number, a confidential message, or secret marketing plans—secure from outsiders' eyes.

11.4.1   Secure Socket Layer Protocol

The Secure Socket Layer protocol was developed by Netscape and is included in most commercial Web servers and supported by most browsers. SSL runs at the transport layer, and therefore application programs [Hypertext Transfer Protocol (HTTP), file transfer protocol (ftp), telnet, etc.] can run on top of it. SSL relies on two interrelated technologies: public-key cryptography for encryption and Digital ID certificates for authentication.

With public-key cryptography enabled, users can transmit credit-card numbers and other data safely—the user feels secure. With Digital ID certificates, the merchant is comfortable that the transmitting user is who he or she claims to be—the merchant feels secure.

To use SSL, the Web server software must be running on a port other than the standard TCP/IP port 80. Secure transactions are usually run on port 443. The secure uniform resource locator (URL) that results from running a Web server on this secure port starts with the access method https:// rather than just http://.

For more information on Secure Socket Layer, including a link to the most recent IETF Internet draft, access Netscape's Web site at www.netscape.com/nevesref/std/SSL.html.

11.4.2 S-HTTP

Secure Hypertext Transfer Protocol (S-HTTP) is server software that provides the ability for "secure" transactions to take place on the Web. S-HTTP is an extension of HTTP for authentication and data encryption between a Web server and a Web browser. S-HTTP encrypts the Web-based traffic between client and server on a page-by-page basis. SSL encrypts more of the protocol stack and works on a connection-by-connection basis.

By adding cryptography to messages at the HTTP application layer, S-HTTP ensures the security of end-to-end transactions. Most server software supports S-HTTP, so individual Web transfers can be encrypted, but currently there aren't many browsers that support S-HTTP. If a Web site is running off an S-HTTP server, users type in shttp:// in the URL section of the browser to enter into the "secured mode."

To view the latest S-HTTP Internet draft from the Internet Engineering Task Force, access ftp://www.ietf.org/internet-drafts/draft.

11.4.3 Private Communications Technology

Microsoft's Private Communications Technology (PCT) encryption offering is an enhanced version of SSL supported by Windows NT. It is a good solution for authentication and encryption routines that complement credit-card-based electronic commerce because it separates authentication from encryption. Applications using PCT for authentication, such as a credit-card validation program, can take advantage of a much stronger security algorithm that is not restricted by the 40-bit key limitation. PCT uses a second key of unlimited length to verify that both parties in an on-line transaction are who they claim to be. A server can listen on the same port for both SSL and PCT handshake initiations and use the same record structures when dealing with the different protocols. In addition, PCT provides records that can be decrypted independently, allowing for secure transmissions across an unreliable transport protocol layer such as UDP (User Datagram Protocol) in packages called datagrams.

Like SSL, PCT is transparent to end users. Users need only enable this protocol from their Web browser's configuration dialog box. As with SSL, URLs starting with https:// are used to access documents secured with PCT.

11.4.4   Secure/Multipurpose Internet Mail Extensions

Secure/Multipurpose Internet Mail Extensions, an RSA-based data-encryption scheme, is quickly becoming the de facto industry standard. S/MIME defines how the contents of a message may be encrypted and how the identity of the sender can be verified with a digital signature. S/MIME messages are encrypted from end to end. The drawback with S/MIME, however, is that it disregards one of the main features of the new Internet Message Access Protocol version 4 (IMAP4): the ability to separate text from file attachments.

Until recently, to decrypt a message, the recipient had to be running the same software as the sender. The new S/MIME standard for Internet E-mail security changes all this, because S/MIME programs are interoperable.

The S/MIME specification is vendor-driven and calls for defining how data can be encrypted or digitally signed for Internet mail content. The body responsible for developing S/MIME as an Internet standard is the Internet Engineering Task Force, and that group is currently reluctant to give S/MIME its blessing, since it uses proprietary technology from RSA Data Security.

A draft of S/MIME 3 was released in mid-1997. New extensions include the addition of signed receipts, labels that identify the sensitivity of messages, and the ability to send encrypted messages to a mailing list. These extensions were drafted by a group that represents the Internet Mail Consortium industry group and government contractors, who know the rigorous security demands of government E-mail networks.

Most major messaging vendors, such as Lotus, Netscape, and Microsoft, have released versions of their products that support S/MIME. RSA Data Security published a description of a key encryption algorithm that provides interoperability between domestic and imported versions of S/MIME-compliant products. But even as S/MIME gains all this market approval, it still does not have an official nod from the Internet Engineering Task Force.

11.4.5   Pretty Good Privacy

Pretty Good Privacy, Inc., is one of the world leaders in digital-privacy software for individuals and businesses. PGPmail 4.5 is the first commercial version of the popular freeware program which provided public-key encryption of E-mail messages. Pretty Good Privacy's core E-mail encryption product was originally developed and distributed as freeware in 1991 by Philip Zimmermann. It seamlessly integrates with the Netscape Navigator Gold mail facility and the popular Eudora mail package.

PGPmail is based on a public-key cryptography system that allows each user to generate a public key (which can be distributed openly) and a private key (which is never shared). The public key is used to encrypt messages to a recipient. The private key is used to decrypt those messages.

The new version of PGPmail will be capable of embedding digital signatures, performing data compression before encryption, and managing public keys, which allow for the distribution of encrypted messages with digital signatures that identify the sender.

11.4.6   Secure Electronic Transactions

Visa and MasterCard, along with software and hardware vendors such as Netscape, Microsoft, IBM, VeriSign, and other major players in Web commerce, cooperated on the Secure Electronic Transactions specification for on-line commerce to take security beyond encryption. SET, which complements the SSL protocol, is a de facto standard protocol for securing on-line credit-card payments via the Internet. Based on public-key-encrypted digital signatures, SET aims to protect transactions and reduce fraud.

Backed by Visa, MasterCard, and numerous major U.S. and international banks, SET creates an infrastructure for developing digital certificates. SET uses software, digital certificates, and encryption technology to enable customers, merchants, and financial institutions to authenticate the identity of parties to a transaction in order to eliminate the risk of fraud.

See Sec. 17.6.1 for more information on SET.

11.4.7   Transaction-Layer Security

Two proposals under review by the Internet Engineering Task Force merge Kerberos with public-key encryption schemes. Kerberos is a "secret key" technology that uses a server to hold all of the keys. Users gain access to resources using a specific Kerberos password that authenticates users by comparing the password to a corresponding key. The two draft recommendations would combine the performance benefits of Kerberos with the security and flexibility of public-key variants such as digital certificates and transport-layer technologies such as SSL.

One such proposal is being worked on as part of a broader effort called Transport Layer Security (TLS), which is likely to become a standard by the beginning of 1998. TLS is based heavily on SSL version 3, but also incorporates Microsoft's PCT. TLS includes enhancements in session authentication, such as the inclusion of digital signature blocks, to replace the traditional site certification in place today.

SSL is based on public/private-key encryption technology, in which a user owns a private key to encrypt messages, which are decrypted by public keys. Under the TLS proposal, SSL would support Kerberos encryption for the first time.

The other proposal before the IETF calls for all public-key schemes to be supported by Kerberos. Under this method, a user could log on to a Kerberos network using a digital certificate. This would solve one of the major problems of public-key technology—that the private key is often tied solely to a single PC. These extensions to Kerberos will enable a user to log into a network from any PC on the intranet by entering a password and have the Kerberos server pass back a private key for use during the session.

11.5   Access Security

An organization's approach to intranet security should be based on the sensitivity of its data. Big companies should start with a firewall server stationed at the gateway point. Software solutions are fine for a small LAN. Tunneling software allows companies to use public Internet lines instead of leased lines to securely connect disparate internal networks.

Three common techniques are available to keep an intranet secure from unauthorized access from the Internet: IPX-to-IP gateways, firewalls and tunneling, and virtual private networks (VPNs). These are discussed in more detail in Chap. 10, "Infrastructure Requirements."

11.5.1   IPX-to-IP Gateways

Since the Internet and intranets use the TCP/IP network protocol for transportation, an organization should avoid loading the TCP/IP protocol on any server that contains private information—use an IPX-to-IP gateway instead.

Unfortunately, while a gateway protects sensitive data and internal servers from intruders, it doesn't allow an organization to publish and protect private company data, which is one of the purposes of an intranet. If users need to access data that is not on an intranet server running TCP/IP, they will have to connect to the network using the IPX protocol through a separate remote access. A gateway is a good solution only if the objective of an organization is to connect a network to the Internet.

11.5.2 Firewalls

As discussed in Sec. 10.4, a firewall is a network node that is set up as a wall to prevent traffic from one segment from crossing over to another without authorization. A firewall, usually a dedicated computer, sits between the internal network and the outside connection (usually a router) to the Internet. The firewall software protects all intranet servers, including Web, ftp, and E-mail servers, on both sides of the connection.

Firewalls can control access using various criteria, such as IP address, Ethernet address, user name and password, or even time of day. Most products include monitoring, logging, and notification features. These features allow an administrator to see when someone is trying to gain access and who that user is.

Application gateways use application proxies. These are programs written for specific Internet services, such as HTTP, that run on a server with two network connections. The proxy acts as a server to the application client and as a client to the application server. Application proxies are generally more secure than packet filters because they evaluate network packets for valid application-specific data. Their network address translation capabilities prevent internal IP addresses from appearing to users outside the internal network. However, because of the double processing, performance suffers.

Most firewall vendors have incorporated additional security technologies into their products. In addition they are beginning to partner with other security vendors to offer complete Internet security solutions. These solutions include encryption, authentication, antivirus protection, protection from misbehaved Java and ActiveX downloads, and even server load balancing.

11.5.3 Proxy Servers

Proxy servers, a type of firewall, allow companies to provide Web access to selected people by preventing unauthorized inbound traffic and restricting downloading by blocking specific Web sites. If a user's browser is configured to access the Web via a proxy server, cached pages are checked by the proxy server each time a page is requested. If the server has a copy of the requested page, that copy is returned to the user. If it does not, it connects to the requested Web site, downloads the page, saves it to its local cache, and passes it on to the user. Consequently, if multiple users ask for the same page, the cached copy is used multiple times. This eliminates the need to access the Internet multiple times, cutting down traffic load and improving response. Because content changes so often, most proxy servers make a quick comparison of the cached page and the Internet page. If the cached page is out-of-date, the proxy server retrieves the new page, caches it, and sends it to the user. If the cached page isn't out-of-date, it is sent to the user.

A proxy server also hides internal network addresses because it converts them to global Internet IP addresses, and it is these addresses that appear on the Internet.

See Sec. 10.4.2 for more information on proxy servers.

11.5.4   Virtual Private Networks

A virtual private network is a network that uses the Internet to expand its own reach—which allows it to include external users such as customers and trading partners—and uses encryption technology to ensure safe arrival of the data. To the user, the intranet feels the same but is more secure, faster, and more efficient.

Many companies look to carriers and Internet service providers (ISPs) to provide a VPN capability from their internal network. However, security and performance are guaranteed only within a particular service provider's network. Once a user hops off one network and onto another—either the public Internet or another private IP service—the only security in place is the encryption used on the message itself.

Currently, there needs to be a firewall at least at each end of a VPN, and the firewalls need to be the same. IETF's IPSec (IP Security) protocol will enable firewalls from different vendors to establish a secure Internet channel—a VPN. IPSec, a part of IPv6, encrypts the tunnels so that privacy and authenticity for an IPX packet running over a VPN, for instance, is ensured. VPN products that support IPSec communicate public keys and encryption algorithms with each other to set up VPN sessions.

IPSec include facilities for encryption, authentication, and key management. However, the protocol adds extension headers to identify encrypted IP packets, and this added overhead may cause a packet's size to exceed the size limit imposed by some routed networks.

See Sec. 10.6 for more information on VPNs.

11.5.5 Tunneling

Encryption encodes data packets for their transmission over a network. A firewall- or PC-based encryption package also scrambles the packets using some sort of encryption algorithm so that they cannot be read if intercepted. The packets are decrypted by the destination firewall or client machine on the other end.

Tunneling is encapsulation—a protocol that surrounds (encapsulates) foreign data packets. IP networks, like VPNs, can handle only IP packets; other types of data, such as IPX, must be packaged inside IP packets. Tunnel software can be used to create a secure intranet through the dial-up network, thus creating a virtual private network. IP tunneling software lets TCP/IP networks carry information packets for the other protocols and guarantees the sender's identity.

Tunneling technologies using two-key encryption systems have been around since the mid-1970s and have been used in client/server configurations. The server side encrypts something that can only be decrypted by the software on the client, and vice versa.

Many VARs rely on browser-based encryption schemes, such as Netscape's SSL, which does encryption at the application level, meaning that only Web information is encrypted. But these solutions are not nearly as comprehensive as tunneling software which works at the IP level. With tunneling, *all* traffic is encrypted—telnet, ftp, gopher, etc.

Virtual private networks require secure or encrypted tunnels. There are two new types of tunneling protocols today that allow remote Point-to-Point Protocol sessions to be accepted by an Internet service provider and authenticated by queries to corporate security databases:

- Point-to-Point Tunneling Protocol, a technique promoted by Microsoft in its Windows NT servers
- Layer 2 Forwarding, used by Cisco in its routers

These two tunneling protocols allow Internet service providers and carriers to receive an incoming call from a remote user and tunnel it directly to a corporate LAN using the Internet. But these are proprietary protocols. Cisco and Microsoft jointly developed the Layer 2 Tunneling Protocol, which allows the authentication and authorization process to be forwarded to a server located on the Internet.

These protocols are discussed in more detail in Sec. 10.5, "Tunneling."

11.6 Hacker Security

Network administrators know that hackers can be relatively benign or do major damage. Security cannot be an afterthought for an intranet, whether it is entirely internal or accessible through the Internet. Organizations must be able to protect their resources from internal as well as external threats.

Officially, there is no way to take an accurate accounting of security break-ins, since most targeted organizations refuse to discuss incidents publicly or even report them to authorities for fear of scaring off their customers or clients. According to a survey of *Fortune* 1000 companies by WarRoom Research LLC in cooperation with the U.S. Senate, 58 percent of the respondents reported that they had experienced a break-in in the previous 12 months, and about 80 percent admitted to having insider threat problems. Most acknowledged that the biggest threat comes from inside the organization, but more and more the damaging attacks are coming from the outside. Over 18 percent had incurred losses of over $1 million each, and only 16 percent actually had staff dedicated to computer/information security.

Type "hack Win95" into most Web search engines—the results will be numerous references to instruction guides, discussion lists, and source code for hacking into Windows 95. The same goes for most current operating systems. As users are provided with on-demand information and computing power, anyone with physical or digital access to a network has the opportunity to hack.

There are diagnostic tools actually being used by hackers to automate their probes for system vulnerabilities. Automated hacking tools can be used to capture and decrypt passwords, dial modems and attempt to log in, and probe systems for security holes.

Most external hackers go on reconnaissance missions before they actually execute a full-blown attack, according to security experts. Sometimes a system administrator or Webmaster detects a hacker casing a site or a network by noticing some suspicious attempts on different ports, such as the File Transfer Protocol, Telnet, and Sendmail, and is able to shut the hacker out before any damage is done.

11.6.1 Password Sniffing

One of the simplest and most common attacks is the password-sniffing method, using tools like TCP Grab and Passfinder, which can be downloaded from the Internet. The password-sniffing program captures passwords, giving the hacker free rein inside the site.

Hackers grab the Packet Handling Function (PHF) program and use it to extract the site's password files, which are then run against a password-sniffing tool to gain access to the server. The PHF is basically an example program of how to write CGI code, such as one that searches a database. Holes like these are exposed because most CGI scripts are written not by security professionals but by graphics and administrative people working on a Web site.

To spot password sniffing, check traffic logs regularly. Use one-time passwords, not reusable ones. Keep on the alert for break-in attempts.

11.6.2   Password Cracking

With password cracking, an intruder steals a password file, decrypts the passwords, and then uses them to gain access to files, applications, and systems. This was discovered as one of the major holes in Windows NT, which has since been plugged.

Password files should not have obvious names and should be protected by a password.

11.6.3   Spoofing

With IP spoofing, a hacker poses as a legitimate host, using a fabricated IP address. That tricks the firewall into letting the intruder into the network. Application-layer firewalls can detect this kind of attack, whereas packet-filtering ones cannot.

An emerging type of spoof attack is Web spoofing, where an attacker sets up a fake Web site to lure users in hopes of stealing their credit card numbers or other information. One hacker set up a site called MICR0S0FT.com, using the number zero in place of the letter O, which many users wouldn't necessarily notice in their Web surfing. Several unknowing users pointed their browsers to the fake site without ever noticing that it was a bogus Web site.

This type of invasion is not destructive. But by filling up someone's disk or flooding a network or host, it can reduce or even deny services. The Computer Emergency Response Team (CERT) calls these "denial of service attacks."

The Secure Socket Layer tool that comes packaged with most Web browsers only makes a secure connection to the bogus site. SSL cannot detect the fake Web site. The main line of defense against Web spoofing is adding authentication software between the client and server so that the client knows for sure that it's at the real www.microsoft.com, and the server knows for sure that the client is who it says it is.

Digital signatures are basically electronic IDs—they include a public key and the name and address of the user, all digitally signed and encrypted with a private key. These IDs are proof of identity and proof that a message wasn't tampered with.

Another solution is to disable JavaScript in browsers on desktops that aren't used for development so that the attacker cannot cover his tracks, as well as making the browser's location line clearly visible.

11.6.4   Spamming

Spamming involves bulk mailings of electronic junk mail. While it doesn't directly affect the security of a system, a spam attack can bring the network performance to a standstill. Spam can suck up bandwidth, clog mailboxes, and force machine upgrades to handle the overloads.

Organizations are using Internet mail servers to help curb E-mail abuses such as spamming and spoofing. They control spamming before it invades the corporate network. These servers can refuse mail from selected sites, block mail from certain addresses, and send an alert to a system administrator if someone is trying to send a message through the server from outside the firewall.

Organizations are beginning to look to the legal system for support in prosecuting those that spam organizations. The spammers cite freedom of speech and civil rights; the recipients cite infringement on their ability to conduct business.

11.6.5   SYN Flood

Using an Internet service provider does not eliminate the security worries. ISPs can have their Internet or IP service disrupted by a flood of phony traffic that clogs the provider's network. The most common is the SYN (synchronized number sequence, a synchronization packet for TCP traffic) Flood attack, which nearly crippled Netcom, Panix, and nearly a dozen other providers in 1996. The good news about SYN Flood and other denial-of-service attacks is that they don't hit the actual systems. They basically wreak havoc on the network, sometimes shutting down the service temporarily.

The trouble with a SYN Flood attack is that there's not a lot that can be done about it once it's under way. When BBN Planet detects spikes in traffic at specific points in time, it responds by putting in filters to block the flood of traffic.

MCI Communications recently announced free software that can track down hackers who flood Internet servers with electronic mail or through other means. The software will alert a network manager when thousands of similar E-mails arrive at once. It will then provide the Internet address or addresses to help find the hacker at the source.

11.6.6   Ping o' Death

The Ping o' Death can crash or reboot a large number of systems by sending a "ping" message of greater than 65,536 bytes, the default size. Organizations can install an operating system patch to block the traffic. Recent updates to router software have also dealt with this problem.

11.6.7   DNS Hijacking

In DNS (Domain Name Service) hijacking, an attacker redirects all queries for one domain, such as microsoft.com, to another site, such as ford.com. It's like swapping phone numbers in the phone book. To safeguard against DNS hijacking, check for any duplicate DNS entries. Eventually, encryption programs and other security parameters will make DNS more secure.

11.6.8   Be Proactive

The main thing is to remember that no network is truly safe. Be aware of the different types of attack that are in vogue, say security experts, and take precautions, such as getting rid of the PHF file and instituting one-time passwords. Keep up with new bug reports and vendor fixes. Also, keep a close eye out for suspicious activity around network ports. Being proactive is the only way to minimize the risk.

If there is suspicious action, contact the authorities, as such activity is a federal offense. Move the sites to a secure hosting service from a different Internet service provider that doesn't provide dial-up access to the Internet. Change the passwords, separating letters with numeric signs to make the passwords more difficult for hackers to crack.

Figure 11-7 summarizes the proactive measures an organization can take against the major hacker threats. Also see the CERT's Web site (ftp://info.cert.org/pub/) for advice about computer security risks and protection.

Figure 11-7
Security Measures
against Hacker
Attacks

Hacker Threat	Defense
Masquerading	Authentication
Eavesdropping	Encryption
Interception	Digital certificates/signatures
Address spoofing	Firewalls
Data manipulation	Encryption
Dictionary attack	Strong passwords
Replay attack	Time-stamp, sequence numbering
Denial of service	Authentication

11.7   Virus Protection

Antivirus software is thought of as a client application, but organizations are beginning to realize that scanning is required at multiple points on a network to prevent data from being destroyed—and to save staff from having to clean machines that become infected. Organizations are scanning for viruses at the firewall, on E-mail servers, and at the desktop. Desktop virus scanners are still the only protection from viruses brought in on floppy disks.

11.8   Tips for a Secure Intranet

Inter- and intradepartmental privileges make intranet security more complex. With all eyes turning to protection from the outside world, organizations forget about prying eyes on the inside of the firewall that might just love to get into some of the company's sensitive records. Internal holes can be breached by wayward curious employees or by disgruntled workers looking for revenge. Many studies show that over 70 percent of people who steal information are insiders, not outsiders. To provide more internal protection, organizations are turning to digital signatures as a way to secure their data and access to applications.

Mission-critical information should receive stringent security measures. SSL-based encryption technologies should be applied to internal traffic and session transmissions. Specific access rights should be assigned to employees, and Web-site managers should have access only to their particular area.

Interoperability among firewalls from different vendors using IPSec will establish secure Internet channels. S/MIME will allow users to encrypt, send, and decrypt messages between multiple vendors' mail systems.

While the integration of security technologies into applications starts at the ground level, the creation of enforceable security procedures must start at the top and include the human element.

Evaluate what needs to be secure and what doesn't. And use fast hardware to host the firewall and tunneling servers.

Authenticate users. Keep mailing lists current.

Audit trails are a must. Review audit trails often to assess use or misuse of the network. Train personnel on what to look for.

Contrary to popular belief, everything does not have to be available to everyone. Set up resources so that they are available only to specific groups of individuals on a need-to-know basis.

Manually monitor the system in real time (or use software to do so). Have a backup plan to move the intranet to a secure location in the event that there is a security breach. Make sure that the users who need the intranet can still get to the information regardless of the breach.

Security is important, but a word of caution is also in order. Security is a two-edged sword. The more "locks" there are, the longer it takes to get through. That's true for both the outsiders you don't want in and the insiders you do want in. The more "locks" there are, the slower the performance. Be sure the security is worth the tradeoffs.

Figure 11-8 lists some of the common solutions for security issues.

Figure 11-8

Six Common Internet
Security Problems and
Their Solutions

Security Issue	Defense
Interception of E-mail	Encrypt E-mail using desktop or server encryption hardware or software. Use digital signatures and certificates to authenticate senders and verify that the E-mail has not been tampered with.
Macro viruses from E-mail attachments	Install an antivirus gateway to filter incoming traffic (E-mail and downloads).
Corporate network intrusion	Protect the perimeter with firewalls. Set up an authentication server on the network if remote users are to have access to sensitive internal data.
Theft or alteration of corporate information	Follow the same procedures as for intrusion. Also use encryption hardware or software to encrypt traffic flowing from office to office across the Internet.
Disruption of network devices and services	Protect the perimeter with firewalls. Give remote users tokens or smart cards. Use an authentication server on the network.
Misbehaving Java and ActiveX applets	Configure firewalls to block Java and ActiveX applets, or install a Java and ActiveX gateway to filter out bad applets.

12

Building Intranet Applications

The ability to deploy i-net technology is a direct result of many technologies working together, reinforced by the use of standards, to provide a flexible environment. Moving to intranets simplifies the development of applications. Programming becomes easier because every user has a common desktop access point (the browser) and a consistent network protocol (TCP/IP).

Deploying new applications and distributing updates is simpler. Changes are made on a server application and relayed to all desktop users rather than requiring IT to change the software running on each user's machine. Training requirements are reduced because each application runs off a familiar interface—the browser.

12.1   Internet vs. Intranet Application Development

Don't be fooled into thinking that an intranet is as easy to plan and implement as the company's Web site for Internet presence was. They are two very different applications which happen to share the same technology.

12.1.1   Process-Focused

An Internet site creates presence on the information highway. Developers are concerned with how the site looks to people who stop by—external perceptions. Intranet applications are usually developed to gain internal efficiencies, usually by cutting costs. This happens when the intranet application eliminates or streamlines a workflow process. Restructuring a business process takes more than just some HTML (Hypertext Markup Language) pages and Java or ActiveX code. The development team must understand the ramifications—technical, structural, cultural, and political—of changing the business process.

If the process isn't completely understood, the resulting application could be ineffective. If the technology isn't understood, the result could be an insecure and unreliable application. If the political and cultural ramifications aren't understood, the result could be an application that isn't used.

12.1.2   Requirements Definition

Most Internet sites are designed based on examples. Developers can look at thousands of sites (including those of their major competitors!) and get a sense of what can be done. There is also a sense of what is needed—something better than their competitors' at a minimum.

Developers of intranet applications don't have that same research capability. Intranets are inside firewalls—only users with valid IDs and passwords can get beyond the firewall to see what the intranet application is all about. This is rapid application development at its best and at its worst. Prototyping begins after only a few requirements have been identified. The requirements continue to be refined as the prototyping cycle (code, review, code, review, etc.) continues. Changes in scope should be expected.

12.1.3   Distributed Management of Content

Internet content is controlled by a single entity. Intranet content can be provided from a wide variety of sources. Using intranet technology, employees can directly manage content. It is important that there are policies, procedures, and standards in place to aid in managing the content. It is also critical that each subsystem mesh well with the intranet as a whole.

Policies and procedures for keeping some information confidential and ensuring that distributed information is reliable are just as critical with the new technology as they were before its use. The policies and procedures should be created before the intranet application is launched.

12.2   Development Tools

The development tools used to build i-net applications are evolving as fast as the technology and its use are. The plus side of this quick evolution is that more robust applications are easier to build and maintain; the downside is that the developers are constantly on a learning curve with the new tools.

When evaluating development tools, most corporate developers agree that writing the HTML layer is the least important aspect of the process. More important are the tools for interfacing front-end Web processes with back-end database activities. Knowing C++, Java, PERL, and CGI (common gateway interface) scripting is critical for establishing high-performance transaction-processing environments.

12.2.1   First Generation

The first generation of application development centered on being able to add HTML tags to files that were then displayed by a browser, as illustrated in Fig. 12-1. HTML could be used to add links to words on that display so that another HTML page would display or an application would open up (called launched) to display a file. Applications that benefited from this structure were static, with predictable behaviors.

These tools have been known as Web authoring tools—familiar names such as FrontPage from Microsoft and PageMill from Abode Systems—and even these tools have developed as Web capabilities have expanded. They have become more feature-oriented as more and more productivity software, such as database tools, word-processing software, and spreadsheet software, allows a user to generate output in HTML format automatically.

Figure 12-1
First-Generation Web
Applications

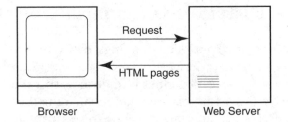

- HTTP listener responds and disconnects
- HTML file system
- URL index
- No real-time updates

Figure 12-2
Second-Generation
Web Applications

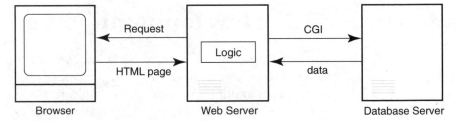

- Web server still has HTML pages
- Application logic in C, C ++, PERL, etc.
- CGI for database interface, reports, etc.

12.2.2 Second Generation

Users quickly wanted more than static pages that were linked together. They wanted to use this type of interface as a way to get data from a variety of sources without having to launch software for each of those sources.

Developers turned to CGI as a way to interact with data-access programs, as illustrated in Fig. 12-2. CGI allows Web-server software to request data from other applications on the server or elsewhere on the network. For example, if a Web page contained the interface to a company's intranet-based employee directory, a user would type a name (or phone extension or office) in a form. CGI would send that data as a request to a database program, which would return the information about that employee as an HTML page, which would then be displayed by the browser.

CGI makes it possible for a customer to track a FexEx package by accessing the FedEx Web page and keying in the tracking number. CGI interfaces format that as a request to a database management system, send it off, and accept the HTML page that contains the information about the package.

Figure 12-3
Third-Generation
Web Applications

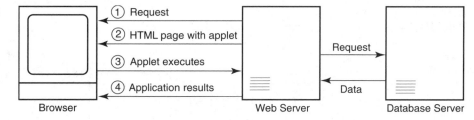

- HTML pages on server have applets
- Client applications downloaded from server and executed on client

12.2.3   Third Generation

Java and ActiveX are the key words for this generation of tools (see Chap. 13 for a discussion of Java and ActiveX). No longer are Web-based applications simply informative; they begin to *do* something. Code [written as a Java program (applet) or an ActiveX component] is maintained on the Web server and downloaded to the client to be run as needed, as illustrated in Fig. 12-3. The downloaded code can be used to generate local reports, perform calculations, and perform validation routines before submitting a form, to name a few possibilities.

Now real applications are possible using the i-net platform. Electronic commerce—on-line shopping—is possible because the order entry system that ran on in-house machines that were accessed by telephone sales personnel is now executed partly on the client machine. The generated shopping cart (list of items to purchase) is electronically fed into the same order entry system.

For example, a customer can inform FedEx of a pickup request by filling in a form on FedEx's Web space. The form gets electronically processed—end result: a truck dispatched to the pickup location. Or a personnel form for change of address is filled out by the employee, validated by downloaded code, and then submitted to the server for processing.

12.2.4   Fourth Generation

The fourth-generation tools use distributed objects to develop applications. These objects can be distributed across multiple computers that can all communicate, regardless of platform, operating system, or programming language—this is distributed computing. These distributed objects can be components of a single application or shared among multiple applications running in the enterprise, as illustrated in Fig. 12-4.

Figure 12-4
Fourth-Generation
Web Applications

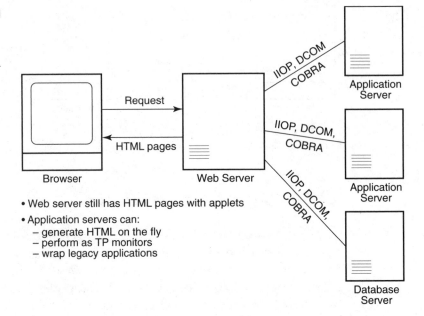

- Web server still has HTML pages with applets
- Application servers can:
  - generate HTML on the fly
  - perform as TP monitors
  - wrap legacy applications

Object request broker (ORB) technology enables distributed computing. CORBA (Common Object Request Broker Architecture) is one of the technologies that is used; the other is DCOM (Distributed Component Object Model). CORBA specifies the communication requirements between the distributed objects. Internet Inter-ORB Protocol (IIOP) focuses on interoperability of distributed objects in heterogeneous environments—CORBA at the wire level.

Traditional Web development is tied to HTTP (Hypertext Transfer Protocol). But HTTP was never designed for application development—it requires that an application send a file to the Web client to interact with the user. This highly inefficient model creates excessive network overhead. Using IIOP, developers can host ORBs on the client or the Web server. The application server's IIOP provides the middleware that enables the application to reach across intranet and Internet boundaries.

IIOP is used by Netscape's Open Network Environment to allow enterprisewide systems to communicate. It is used by Enterprise JavaBeans to allow developers to build scalable business applications with reusable server components. IIOP allows programmers to connect Java, JavaScript, and C or C++ code to enterprisewide systems. It's IIOP within Netscape Navigator's LiveConnect that lets applets, plug-ins, and scripts communicate with one another.

See Sec. 13.1, "Component Software," for more information on object request broker technology, CORBA, IIOP, and DCOM.

12.3 Deployment Issues

Just as client/server is not a technology to rush into, intranet use and development benefits from careful thought and planning. Granted, building a corporate intranet and writing applications as browser-based applets with WWW connections is relatively easy and—to the IT staff—fun. But i-net applications are not a panacea for the problems of corporate application deployment.

12.3.1 Infrastructure Issues

I-net applications can have a great impact on the corporate infrastructure.

- Few people have much experience building or running corporate-wide Web sites. Those who do are probably not in-house, so existing staff is on a learning curve.

- Web sites and intranets take on a life of their own once created—these operating costs have to be factored in.

- As usage escalates, the existing WAN capacity may not be able to effectively handle the additional traffic loads.

- Permitting remote access—by authorized employees, trading partners, or the public via the Internet—may expose a company's network (and any node and data on it) to unauthorized access.

12.3.2 Development Issues

It's easy to get excited about this new application deployment approach. However, the fact of the matter is that the fundamentals of systems development still apply. As is true of any application development, more time is spent assessing needs than in actually producing a site.

Organizations building intranets need to keep in mind that doing the job right the first time is more important than doing the job first. When developing and deploying an intranet, network administrators need to realize that there is no single strategy to employ; rather, there are a series of strategies to consider. The first step in any intranet deployment is to get the various departments that will be affected by it involved and avoid alienating one group from the project. Intranets should not be designed around technology, but rather around business goals. Information departments need to define the goal of the intranet, coordinate its development, and recruit members to support it once it becomes deployed.

The issue of ownership is often overlooked. Everyone wants to be part of the project when it is new, exciting, and "cool," but once it's developed, day-to-day operations are not so exciting. Content is the major ball that gets dropped. How does an organization manage content—and who is responsible for it?

12.3.3   Network Issues

If the organization is not running TCP/IP, there is major work ahead. If TCP/IP is in place, the organization needs to identify and verify how LAN segments are already interconnected if the intranet is to be enterprisewide or even multidepartmental.

It's extremely important to review TCP/IP addresses for each of the network segments that will be used to access the intranet servers. IP brings some address and subnetwork issues along with it. Many organizations have fallen into the trap of simply creating their own IP naming schemes—sometimes allocating Class B or Class C IP addresses that have not been registered with the InterNIC. In some instances, the same IP addressing scheme has been used on different LANs that have no connection to one another. As soon as a router connects them, or if one is connected to the outside world, these networks could stop dead.

In the 1980s, departments learned how worthwhile and easy it was to create their own networks. However, decentralized networks led to inconsistent use of procedures such as naming conventions, protocols, services, and server types. When organizations tried to integrate these LANs on an enterprisewide scale, they ended up with chaos. Departments that didn't like what IT was selling—standardization—just didn't join in.

12.4   Deployment Hints

The benefits of intranets are becoming apparent to corporations, but the cost savings may be only short-term as organizations begin using intranets for more and more enterprise tasks. When deploying an intranet, an organization should implement it as part of the overall IT budget and not create a special project category for it. The organization should start the intranet by simply publishing information, and establish policies for intranet use early on. Intranet applications should be rolled out fairly quickly, and the technology should be deemphasized.

Development of a comprehensive, efficient intranet involves a few simple principles—most of which usually apply to the initial use of any new technology.

- Developers need a real understanding and appreciation of intranet technology to maintain enthusiasm through the implementation process. Their enthusiasm will help convince the skeptics of the worthiness of the project.

- Primary business areas that can benefit from an intranet should be identified. Areas should be identified for both the pilot and the next stage of the technology rollout. The use of the intranet in these areas has to make a difference—go for the "Wow, this is great!" reaction rather than the "This is nice" reaction.

- Ownership must be identified. Who is responsible for the site's content? Who is responsible for technical support of the site?

- Senior management support, as well as support from business users, is crucial to the success of the project.

- Network managers should realize that intranets will not lessen their workload, but will simply shift the burden.

- Sufficient bandwidth, domain name services, security, and backup procedures are the backbone of an intranet strategy. Can the network sustain the traffic? Is bandwidth adequate to ensure reasonable response time? Will the security prevent user access to confidential material?

- Identify alternate access methods. Is there a non-Web-enabled route to get to mission-critical data for those that can't access the Web?

- Demonstrate the utility of intranet technologies in the business environment for areas outside the pilot—generate excitement. Get users involved as developers.

As basic content and design tools become more widely available, cheaper, and more feature-filled by the day (or so it seems), time is much more critical than dollars. Be prepared to outsource some tasks—for example, hiring a professional graphic artist to create and maintain those nifty graphics everyone wants.

IBM is offering e-business Advisor, a series of reports and tools for building an intranet, on its Web site for free (at the time of writing). The service presents 10 scenarios, ranging from just considering the construction of an intranet to setting up the schedule to implementing the final plan. Users start with the intranet implementation planner, where they can answer some 25 questions, shortly after which a plan will be generated. Among the other e-business Advisor features are progress indicators that assess the level of difficulty that will be faced and a text summary that indicates the progress of the planning process and the resources available to complete the site. The text summary also invokes the intranet function design to determine the site's functionality.

12.5 Deployment Strategies

Organizations benefit from intranets in a variety of ways that cannot be directly measured, and it may be difficult to win over the skeptics at first. Intranets will be successful if end users can contribute content.

While having a policy governing who can publish content and how they do it seems to be straightforward, questions surrounding who should have access to information usually are not. Not all content should be available for all employees to see, and proper measures should be taken to restrict access appropriately.

However, the flip side of that—ensuring that users use Web applications in the manner intended—is equally challenging. Many IT managers fear that users will spend their time surfing instead of working—so much for productivity gains. If it is easy for users to check their 401(K)s, they will.

Access always seems to be such a big issue. The ironic thing is that when deciding on policies that address access to the intranet, organizations are just implementing traditional information policies—only in a different way. The intranet makes it easier to share information but doesn't change who should see it.

For example, a company's human resources home page could allow managers to view employees' salary information—they had access before the intranet, just in a different form (probably printed). The policy that other employees should not see salary information has not changed. Intranets simply need to enforce the existing policy.

The following security mechanisms are commonly used to provide different levels of access to corporate intranets (and also discussed in Chap. 11):

- *Login and Password Authentication.* Users can be required to log in before gaining access to a given Web application. This login protection can apply to a set of pages within an application (no access to any salary pages, for example) or to other services, such as to a database server containing sensitive data.

- *Encryption.* Since it is relatively easy to peek at information traveling on a network, encryption technology is worth considering. For example, Secure Socket Layer (SSL) and Secure HTTP (S-HTTP) are two common approaches to encryption that can provide protection at the directory or page level. The bigger question is what to encrypt and what not to encrypt.

- *Firewalls.* Most organizations put a firewall between the company and the Internet. Many organizations are also employing firewalls inside corporate walls. For example, an accounting department could isolate itself from the rest of the company by a firewall on its LAN— and still have access to other LANs. In addition, the firewall would allow internal authorized users to pass through to the accounting department's LAN and thereby to its applications and data.

12.6   Outsource Intranet Development

In many organizations, time is becoming more critical than dollar resources. Intranets can affect the organization's bottom line. They can open up communication among its employees. But first, someone has to build the intranet, upgrade the network, and decide on and organize the content. Intranets take weeks, if not months, to design and launch. Organizations need to choose their technologies (the easy part) and deal with some thorny organization issues (the hard part).

12.6.1   Consultancies

Organizations are turning to intranet consulting firms. These include familiar names such as Cambridge Technology Partners, Sapient Corp., Hewlett-Packard Co., IBM, Sun Microsystems Inc., and Digital Equipment Corp. There are also a lot of newcomers and scores of small businesses. Some are essentially graphics shops that have grown from HTML page design to building intranets, and some are intranet-consulting startups.

The skills these consultancies bring to the table go beyond HTML, page design, and technological expertise. They help an organization derive a comprehensive view of integrating and managing intranet applications. Intranets tend to be departmental rather than the enterprise solutions Netscape is talking about. These consultancies can assist the IT staff in learning the new technology as the project grows. They understand Internet technologies and can go beyond the hubs and routers that LAN administrators are focusing on. They know what intranet designs scale well.

Consultants can be the referees among competing departments as well as between IT and those departments. The responsibility for the intranet's content needs to be assigned. The ownership—and what that means—needs to be understood and assigned.

CompuServe, Cable and Wireless, Equant International, Infonet Services, and IBM Global Network are all offering to handle outsourced global intranets, including servers, infrastructure, IP, Web hosting, and more. However, caution must be exercised when outsourcing a global intranet. Vendors can extend their networks by sharing POPs (points of presence), but at the risk of degraded performance. Coverage is patchy, especially outside North America and Europe. Network managers need to still handle such chores as handing out Web addresses and installing Web browsers.

12.6.2   Value-Added Resellers and Integrators

Companies contract with value-added resellers (VARs) and integrators for intranet-development services because these firms have not only basic Web-development expertise but also integrated solutions to some of the thorniest tasks facing corporate developers. They offer custom software solutions for Internet security, dial-in access, fax broadcasting, chat/conferencing systems, and on-line transactions—all bundled together.

Some companies turn to VARs and integrators to jump-start intranet projects because of their targeted market expertise. VARs can come in with a more seasoned perspective than the internal IT folks have, since they've seen how a number of companies and organizations have solved similar problems.

Integrators and VARs can bring highly targeted expertise to customers in their chosen vertical markets because they are in the trenches. They see what's on the leading edge. They develop a workable understanding of "best practice" within their market domains. After working with a number of companies to devise an innovative approach to a particular problem, the integrators and VARs develop a very broad, very knowledgeable perspective on that problem. It is rare that customers would arrive at that same level of understanding on their own as quickly.

Most VARs say that the biggest issue they face when implementing intranets is security. The second-biggest concern is content management.

Some feel that effectively linking an intranet with legacy applications and databases is the best way to simplify content management. By leveraging what's already there, an organization can establish an intranet that dynamically updates the Web site based on changes to the underlying databases. Because many of the legacy systems being accessed by the Web applications are proprietary, HTML isn't always easily integrated with these systems. A Web-development environment that is open is needed to handle this integration.

Another big issue that customers face is deciding how to manage their intranets after the developers are gone. Content management is key to the success of the intranet. There are very few tools available to help with that management. The organization has to be willing to invest time and resources to make sure that content management happens, and in a timely fashion.

13

Software for Development

The current capabilities of i-net technologies are a direct result of the maturing of many technologies, the most important of these being component software.

13.1   Component Software

Object technology is at the foundation of i-net development. Objects are independent program modules that are designed to work together at run time without any prior linking or precompilation of the group—the software version of "plug and play."

With objects, a system can be designed as familiar business functions and the design carried down to the programming level. In traditional systems analysis, programs are decomposed into procedures, which usually do not fit the business model. Because of the paradigm shift, object technology can be more easily understood by non-computer professionals who have not been immersed in procedural logic.

ActiveX and Java are both based on object technology. An object is a set of functions collected into interfaces. Each object has data associated with it. The source of the data itself is called the data object.

What has risen out of object technology is component software. These architectures provide the interfaces that let independent program modules communicate with each other at run time. Components can be large or small and can be written in different environments, which makes them platform-independent. The terms *object* and *component* are often used interchangeably.

13.1.1   What Is a Component?

A component is a reusable software module—a collection of business objects that handles processing, encapsulates data, and provides necessary user interfaces. A component may look and feel like an object but not fit the academic definition of an object. The following characteristics differentiate a component from an object:

- A component is designed to be used within another application, which is called a container. For example, a container could be a Visual Basic program or a browser. However, a component can also run as a distinct process.

- A component may consist of only one class, be a composite of multiple classes, or be an entire application.

- A component is designed to be reused without modifications to its source code. For an object to be used, its source code must be accessible.

A component could be a simple widget in an application development tool—a development component. Some components are designed to be used within a compound document framework [Channel Definition Format (CDF) components]. Components are also designed to be used to provide a service or application (service components).

Microsoft's Component Object Model (COM) supports development components (ActiveX controls), CDF components (DocObjects), and service components (Automation and native COM objects).

Java applets (CDF) and servlets (service) are specialized components designed to execute within an HTML (Hypertext Markup Language) environment. JavaBeans and Enterprise JavaBeans provide a more generic model for development components and service components, respectively.

A component can't execute on its own. In a client/component model, the client application is designed as the container that holds the other components. The client container is responsible for the user interface and for coordinating mouse clicks and other inputs to all the components. Think of it as the operating system for the component.

The use of components allows huge, feature-rich applications to be dynamically assembled as features are required. Users will call in additional features only when needed. Such features are delivered as applets, small applications that perform specific tasks. An applet could be the Cardfile from Windows, a complete application, a utility program, a limited-function spreadsheet, or a database query. The term was first used with Java programs because Java applets are relatively small in size. It has come to mean any small program that is downloaded over the Web.

Components are commonly used today on clients. They are also beginning to appear on servers. For example, Microsoft Transaction Server acts as a container for ActiveX components participating in transactions.

13.1.2 Object Request Broker

Component software is based on an object request broker (ORB) architecture. Objects within a component use the ORB to interoperate. The ORB also lets the objects exchange information with one another, which allows objects to learn about one another at run time. ORB is a newer technique for enabling objects to extend messages within a single system or across distributed computers. The object orientation provides a higher level of abstraction for connecting objects than an RPC (remote procedure call), messaging, or transaction manager approach.

ORB architecture consists of three components: the object model, the object implementation, and the object request broker itself, as illustrated in Fig. 13-1. An ORB takes a message from a client program, locates the target object class, finds the object occurrence, performs the necessary translation

Figure 13-1
Components of ORB
Architecture

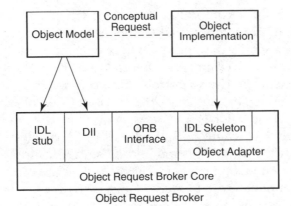

to allow the invoking and called objects to communicate, and passes back the result. The underlying connections are usually performed by a synchronous RPC or messaging system.

Once a client makes a request, the ORB is responsible for all the tasks required to find the object implementation in the network, prepare it to receive the request, and transmit the request. It is also responsible for communicating the results to the requester.

An object contains both code and data. The functions that an object can perform are defined by its interface. Programs that need the services of an object call those services through the object's interface. Consequently, the internal code and data of an object cannot be accessed by any process except the object's services themselves. If an object is changed but its interface remains the same, the changes will have no effect on the programs that invoke the object.

Interfaces are defined using an interface definition language (IDL). The invoking applications make use of the interface information to access local or remote objects, either statically or dynamically. The object interactions are managed by the ORB.

The object model is the architecture definition of the objects in a given system. It also contains code and data that are accessible via interfaces defined by an IDL. The interface definition provides information about the operations that are permitted on that object, the arguments each operation expects, what each operation returns, and what happens when errors (known as exceptions) occur. Programs never have access to the object itself, only to the interface.

Interfaces are stored in an Interface Repository, which provides information regarding interfaces and types at run time. Clients can construct dynamic requests at run time using the Dynamic Invocation Interface (DII). Clients interface with IDL stubs and the DII via a programming language.

A client process issues a request to an object by using an object reference to the object model. The client process does not need to know where the object is or the current state of the object, such as whether it is currently invoked by another process. The object implementation actually executes the services defined by the interfaces and requested by the clients.

The ORB provides the infrastructure needed to deliver requests and their parameters to objects and return the results to the clients. In addition to being able to send requests, receive responses, and deliver messages, the ORB must be able to provide connection management functions and locate objects within the network.

For this to happen, the IDL interface descriptions are compiled to generate a stub and skeleton C++ code that is executed at run time when the object is invoked. Using an object reference, a client can issue requests to the object by making a method invocation, using a stub, or dynamically composing a request using a DII.

Regardless of how the request is made, the ORB core (the portion of the ORB that is responsible for communication) locates the object, establishes the connection to the object, and delivers the request and all related data from the client to the object implementation. The Basic Object Adapter works with the ORB core to locate the actual object implementation and deliver the request in the form of a method invocation to the skeleton C++ code. The object implementation fulfills the request, and the ORB sends a response back to the client if appropriate. Responses follow the same path as the request.

Some industry watchers predict that distributed objects (objects located on different machines and possibly using different object models) is the way applications will be built in the future. But for now, the technology is still in its infancy, with many technological problems yet to be solved.

One problem is how to ensure that the object connections work well and do not overload the network.

Another problem is how to make objects that use different models work together. Common Object Request Broker Architecture (CORBA) lets developers write objects in multiple languages; the objects can then communicate through various object request brokers on machines that run different operating systems. That kind of platform independence is important for high-volume transaction processing, which may require a combination of powerful midrange and mainframe computers that run different operating systems.

Microsoft's Distributed Component Object Model (DCOM), another architecture that allows objects to communicate, currently runs only on Windows NT and Windows 95. Products in Microsoft's Active Platform (Active Platform combines ActiveX and Active Server and connects them with DCOM to create distributed applications) don't support CORBA.

The proposed CORBA 2 addresses interoperability among CORBA-compliant ORBs only, which eliminates non-CORBA-compliant objects such as Microsoft's COM. Different object models do not understand each other's messages or data types.

Using distributed objects does not address the need for transactional support and integrity. Either the ORB or the transaction manager must provide these functions.

13.1.3 Common Object Request Broker Architecture

Most of the work on object-oriented standards has been spearheaded by the Object Management Group. Its specifications for CORBA are used within the industry as "the" definition of an ORB, and the two terms are used interchangeably.

The purpose of CORBA is to provide portability and interoperability of objects across heterogeneous systems. CORBA defines interfaces to various things, including

- The ORB
- Application objects
- Object services

As illustrated in Fig. 13-2, CORBA provides the switching mechanism for passing requests and results between objects. This interoperability becomes more critical as organizations want to bring different platforms, databases, legacy applications, and ORB implementations together in an environment that will support the entire enterprise.

CORBA-compliant ORBs also provide

- *Name services,* which map requesters to methods through some type of object location service
- *Request dispatch,* which determines which method should be invoked
- *Parameter encoding,* which maps the requester's local parameter formats to the recipient's formats

Figure 13-2
CORBA Model That
Uses an Interface
Repository

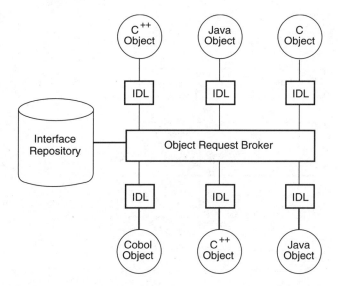

Figure 13-2
CORBA Model That
Uses an Interface
Repository

- *Delivery,* which can deliver requests and results using protocols and ISO standards
- *Synchronization,* which provides a means for a recipient to reply to a requester in a timely manner
- *Activation,* which actually invokes the method
- *Exception handling,* which provides restart/recovery resources
- *Security,* which provides authentication and protection

The CORBA technology and ORB implementations can be used to link legacy applications to client/server applications. An IDL (interface definition language) interface for the legacy application turns it into an ORB object. This interface is called a wrapper. The legacy application is then accessible by any ORB client. The wrapper does not modify or disrupt the existing legacy application. If additional functions of the legacy application must be used by other applications, the wrapper is modified to reflect the new service.

13.1.4   Component Object Model

The Component Object Model from Microsoft is an object system that prescribes its own model. CORBA is an object model. COM is the underlying object model and implementation for OLE (Object Linking and Embedding) and is used in desktop applications to provide a standard for software-com-

Figure 13-3
COM Architecture

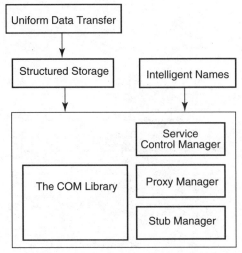

The Component Object Model

Figure 13-4
Accessing COM Services on the Same Machine

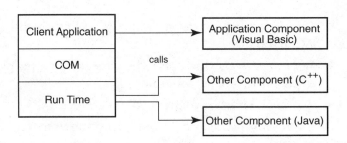

ponent interoperability. The COM architecture is illustrated in Fig. 13-3. ORBs are used as the infrastructure to construct enterprise-class distributed systems. [Not to be left out, IBM's OpenDoc desktop component model is based on its System Object Model (SOM), which is based on CORBA.]

How COM services are accessed is illustrated in Fig. 13-4. The COM and CORBA object models differ substantially. The CORBA model is the most familiar to object developers. Its interfaces have methods and properties associated with them and may be defined utilizing multiple inheritance. COM, and therefore OLE, uses interfaces and classes differently. The component object can have more than one interface, although a program can have a handle to only a single interface. COM interfaces typically have no inheritance, and COM itself can support only single inheritance.

While COM has a basic interface, most COM objects are accessed through a simpler, dynamic mechanism called Automation. Automation supports simpler data types and a simpler invocation process, and can be used with a wider variety of programming languages and applications.

However, given their platform preferences, it is likely that COM will continue to dominate the desktop and CORBA will continue to dominate in the enterprise.

13.2 Distributed Components

When components are distributed between servers, an ORB is put between them, which allows them to connect. When this distribution is over the Internet, special protocols are used.

13.2.1 Internet Inter-ORB Protocol

Internet Inter-ORB Protocol (IIOP) is the CORBA message protocol used on the Internet. CORBA allows programs (objects) to be run remotely in a network—distributed objects. IIOP links CORBA's General Inter-ORB protocol (GIOP), which specifies how CORBA's ORBs communicate with each other, to TCP/IP, the Internet's transport protocol.

Messages are used to request remote method invocations on CORBA objects. IIOP specifies a set of message types that an ORB must support and defines a common data representation for the message content.

IIOP is built into Netscape Communicator 4.0, which is expected to give a tremendous boost to the use of CORBA objects over the Internet. When a user accesses a Web page that uses a CORBA object, a small Java applet is downloaded into Netscape Communicator, which invokes the ORB to pass data to the object, execute the object, and get the results back.

Sun uses CORBA and IIOP to implement heterogeneous remote procedure calls in Java 1.1. The Enterprise JavaBeans API (application programming interface) uses CORBA and IIOP.

13.2.2 Distributed Component Object Model

Distributed COM is sometimes called COM with a longer wire. It allows desktop applications to work with remote components across various kinds of network connections, using the familiar COM APIs that developers are already using in component-based Windows applications. However, Distributed COM makes a network appear to any given application as if it's a local environment—it uses the PC model of a single user as a the central point of control.

Figure 13-5
Accessing COM Ser-
vices on Multiple
Machines (DCOM)

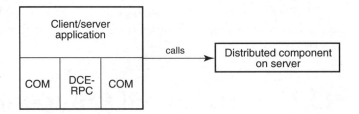

Distributed COM is designed for use across multiple network transports, including TCP/IP and NetBIOS. How COM services are accessed
across multiple machines (or Distributed COM) is illustrated in Fig. 13-5.

Distributed COM is Microsoft's alternative to CORBA. COM was
designed to support communication between Windows applications on a
single desktop machine. DCOM enables communication between COM
objects over a network. In 1996 Microsoft extended COM and DCOM to
provide Web support and named the program ActiveX.

13.2.3   Remote Method Invocation

Sun's Remote Method Invocation (RMI) technology lets Java programs
invoke object methods across the wire using a lightweight protocol.

RMI currently relies on JavaSoft's own RPC protocol, which will ultimately be replaced with IIOP—JavaSoft will, at that point, abandon the
proprietary ORB on which RMI is currently built. Sun is making the
move in recognition of the fact that the world is not all Java. RMI is Java-
centric; it focuses on the communication between Java objects. IIOP is
generic; it allows communication between all kinds of objects.

13.3   ActiveX

ActiveX is a brand name, not a precise technology label. It includes

■ A language-independent object model—COM

■ Conventions for component software—ActiveX controls/components

■ Distributed object support—Distributed COM

ActiveX (formerly known as Object Linking and Embedding) can be
used to add functionality to static HTML pages. ActiveX enables components to interact with one another.

ActiveX is built on the Component Object Model. COM is a language-
independent component architecture that provides a platform-indepen-

dent and distributed platform for multithreaded applications. COM lets components communicate with each other. Distributed COM lets them communicate over the network. Distributed COM is similar to CORBA's Object Request Broker.

COM also encompasses everything previously known as OLE Automation, which let higher-level programming languages access COM objects. With COM, the transfer of the object data itself is separated from the transfer protocol.

13.3.1   Elements of ActiveX

The elements of ActiveX are

- ActiveX controls
- ActiveX documents
- Active scripting
- Java Virtual Machine
- ActiveX Server Framework

ActiveX Controls   ActiveX controls are the interactive objects in a Web page that provide interactive and user-controllable functions. ActiveX controls [actually OLE controls (OCXs) that have been specialized to support the Web] add behavior to Web pages. ActiveX controls can reside locally on a client machine, or they can be downloaded from the Internet or an intranet. An ActiveX control can be as simple as a button or drop-down list or as complex as a query tool.

ActiveX Documents   ActiveX documents enable users to view non-HTML documents through a Web browser. ActiveX documents automatically link documents in a variety of formats to the Web site. Using a consistent interface, users can view and edit documents that can be referenced with HTML (such as a Microsoft Word document). Internet Explorer 4.0 allows users to browse their local hard disk, their LAN drives, or the Internet from within the same user interface.

Active Scripting   Active scripting controls the integrated behavior of server ActiveX controls and/or Java applets from the browser or server. The current Active Script language is VBScript, which is a streamlined subset of Microsoft Visual Basic and is easy to use even for non-VB programmers. VBScript offers direct support for the ActiveX and Active Platform technologies. VBScript is ideal for working with text-based files, numbers, and strings. The language can also be used to create forms and controls for manipulating Web-page elements.

Java Virtual Machine   The Java Virtual Machine (Java VM) is the code that enables any ActiveX-supported browser, such as Internet Explorer, to run Java applets and to integrate Java applets with ActiveX controls.

ActiveX Server Framework   The ActiveX Server Framework provides a number of Web-server-based functions, such as security, database access, and others.

13.3.2   ActiveX Controls

ActiveX technology is based on the notion of component applications. ActiveX controls are downloaded to the client machine once, and then used by any number of applications. ActiveX controls are language-independent and can be developed using C++, Visual Basic, or Java. ActiveX objects are automatically downloaded if the object is not already in the user's machine. An ActiveX object can be as trivial as a single button for a graphical interface or as complex as a complete application.

ActiveX controls are interactive objects created by programmers. They can be embedded in documents, in programs, and on Web pages. They are language-independent and can be programmed in languages such as Visual Basic, C++, and Java. Over 1000 ActiveX controls are available today, including the Macromedia Shockwave for Director control and the Adobe Acrobat control.

Think of ActiveX controls as self-installing plug-ins for Windows-based systems. If an ActiveX control is to be used, it must be registered with the operating system. A control automatically enters information about its existence in the Windows registry so that OLE container applications know how to use the control. When Internet Explorer 3.0 (or higher) encounters a Web page with an ActiveX control (or more than one), it checks the user's local system registry to determine if the component is already available on that machine. If it is, Internet Explorer displays the Web page and activates the control. If the control is not already installed on the user's computer, Internet Explorer finds and installs the component from the Web, based on the location specified by the developer creating the Web page. The downloading service supports versioning, so new versions of the control can be detected and downloaded as necessary.

Many of Microsoft's offerings can use ActiveX components. Applications can call objects and also expose their own internal functions as objects. For example, a spreadsheet ActiveX client could use the ActiveX objects in a database program to retrieve data from the database and store it in spreadsheet cells.

ActiveX controls also include a licensing mechanism to prevent the unlicensed use of controls that are downloaded to support individual Web pages but should not be used for development. Controls are distributed with either a developer's license or a run-time license.

Microsoft is also providing certification services using digital signatures to guarantee that controls distributed over the Internet arrive uncorrupted by Web-based intruders. When a file or ActiveX page is downloaded, a check is performed to determine whether the file or content was signed by its publisher and whether it is valid.

Microsoft offers the ActiveX Control Pad as a development tool for inserting ActiveX controls into HTML Web pages. A free beta copy of the ActiveX Control Pad can be downloaded from Microsoft's Web pages (at the time of writing).

13.3.3   ActiveX Containers

ActiveX containers can run OCX (OLE Custom controls) and ActiveX components. OCX is Microsoft's second-generation component architecture. Visual Basic controls (VBX) were the first, and ActiveX is the third.

13.3.4   Automation

Microsoft uses the term *automation* for executing objects. *ActiveX automation* refers to that part of ActiveX that allows a client program to dynamically invoke the processing in an ActiveX object.

13.3.5   Active Desktop

In a totally ActiveX world, clients run Active Desktop, where embedded ActiveX controls provide the user interface to remote services. These components send requests via either HTTP or COM to a middle-tier application server. There, Active Server Pages may use server-side Visual Basic scripts to query an SQL database via ActiveX Data Objects. These generate Dynamic HTML pages for returning the query results to the client and call server-side ActiveX components running under Microsoft Transaction Server (MTS) that perform any application processing.

MTS provides an environment for executing Distributed COM built from ActiveX components communicating with one another via the COM protocols. MTS handles all the management of sharing, processes, and threads. All the components that make up an application can share these resource pools.

13.4   Java

Java from Javasoft, a division of SunSoft Inc., is a programming language from JavaSoft, a division of Sun Microsystems. Java was modeled after C++, and Java programs are called from within HTML documents. Java was designed to run in small amounts of memory and provides its own memory management. Java is basically a virtual machine and interpretive dynamic language. Because the interface is separate from the implementation, Java applets can run transparently anywhere in the enterprise.

Any Java application can run on any system that supports a Java platform; that means that Java applications are completely portable. Java platforms have been implemented on nearly every operating system and hardware platform available. Subsets of the Java platform are available that run on smart cards (JavaCard), embedded microprocessors (Embedded Java), and personal electronic devices (Personal Java).

13.4.1   Java Platform

Java is used to write applets (small self-contained modules of code). Each applet is embedded in (attached to) a Web page. Java applets work within a Web page but, unlike ActiveX controls, can't access local files. Applets are not persistent—the applet is downloaded each time the Web page is. When the Web page is released, the applet is also released.

When the Web page is downloaded, the applet's Java code is downloaded as byte code. This intermediate language is then executed by a run-time interpreter which translates it into machine code and runs it. Because of this, Java applets are not dependent on any specific hardware and will run in any computer with the Java Virtual Machine. Java applets can be compiled into machine language to be run on a server but lose their portability.

Since Java is an interpreted language, Java applications require an interpreter run-time environment referred to as a Java platform. A Java platform consists of a Java VM, a Java Class Loader (a simple ORB), and a Java Security Manager. The Java Security Manager ensures the integrity of a Java applet or application. It ensures that applets are not corrupted during transmissions, handles authentication, and ensures that a Java applet or application behaves in a friendly manner.

13.4.2 Java Virtual Machine

Java is currently royalty-free to developers writing applications. However, the Java Virtual Machine (Java VM), which executes Java applications, is licensed to companies that wish to incorporate it into their browsers and servers. The Java VM is a Java interpreter from Sun. It converts the Java byte code into machine language one line at a time and executes it. A just-in-time (JIT) compiler can dramatically improve the performance of a Java VM. A JIT compiler is called as a Java application is loaded into the Java VM. The Java byte code is dynamically compiled and converted into platform-specific binary code at load time.

Microsoft also calls its Java interpreter a Java Virtual Machine. The Microsoft Java Virtual Machine is the code that enables any ActiveX-supported browser such as Microsoft Internet Explorer to run Java applets and to integrate Java applets with ActiveX controls.

Java is the programming language, and HotJava is the software on the desktop inside the browser that runs the Java applets. HotJava executes Java programs embedded directly within Web documents.

13.4.3 JavaScript

Because Java is a simplified subset of the C++ object-oriented language, it eliminates some complexities and redundancies. Java supports classes, inheritance, methods, polymorphism, dynamic binding, and other constructs.

Java is a programming language that is not intended to be used by the casual programmer or the end user. JavaScript from Netscape is somewhat easier to use but not as powerful. JavaScript, a scripting language reminiscent of Basic, uses the HTML page as its user interface. On the client, JavaScript applets are maintained in source code. On the server, they are compiled into byte code like that of Java programs. If the applets are expected to be run off the server, they are called servlets. Servlets are roughly equivalent to Microsoft Active Server Pages.

JavaScript is used in conjunction with Java. JavaScript is used to tie Java applets together. For example, a JavaScript applet could be used to display a data entry form and validate input, while a Java program processes the information.

Unlike VBScript, JavaScript is completely object-based.

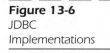

Figure 13-6
JDBC
Implementations

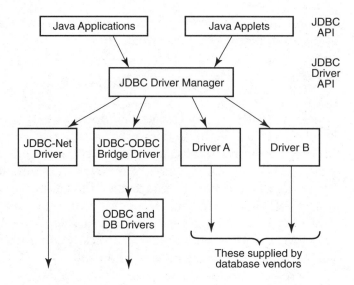

13.4.4 Data Access Using Java

Java is not database-specific, but it can be applied to database problems.

JavaSoft has developed the Java Database Connectivity (JDBC) API specifically to address database access. The JDBC architecture is illustrated in Fig. 13-6. This is a set of Java Core Classes that define a standard mechanism to access and manipulate a database. JDBC is roughly based on Microsoft's Open Database Connectivity (ODBC) standard. Before JDBC, database-access solutions were specific to a particular database. But because JDBC is implemented on top of several common SQL-level APIs, including ODBC, it provides a common API to all major relational databases, and even some legacy ones. As a result, JDBC is enjoying hype and popularity at Java-like levels.

13.4.5 JavaBeans

Java applets provide a single component model; they can't interact with one another. JavaBeans provide a model for building Java components that can be dynamic and interactive.

JavaBeans are development components that can be used with any Java application development tool. JavaBeans are independent Java program modules that work together at run time—they can be dropped into an application container. A JavaBean is analogous to an ActiveX control used

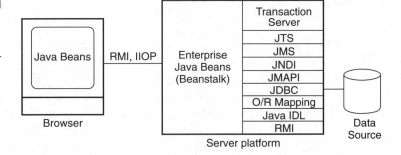

Figure 13-7
Java Platform for the
Enterprise

in Visual Basic. JavaBeans components can be run remotely in a distributed computing environment using Sun's Remote Method Invocation (RMI) or OMG's CORBA. Using RMI calls, an applet running on a client can control a servlet more directly than is possible using HTTP.

JavaBeans are components within the JavaSoft Enterprise Framework, as illustrated in Fig. 13-7. Current Java development requires developers to interact with class libraries to build applications. JavaBeans, in contrast, contain the core business logic of enterprise applications and provide a higher-level interface to the libraries. Consequently, developers can quickly develop applications by linking JavaBean components.

13.4.6   Enterprise JavaBeans

Enterprise JavaBean architecture looks similar to ActiveX but for two differences:

- JavaSoft doesn't supply a full transaction-server environment. It is relying on names like IBM and BEA to do that.
- JavaSoft is staying with its operating system–independent architecture, relying on the cross-platform nature of the Java VM.

Enterprise JavaBeans are an extension to the JavaBeans model targeted to the middle tier. The lightweight JavaBean model is extended to include multiuser security and resource management features similar to those of ActiveX. The specification for Enterprise JavaBeans was released as beta at the end of 1997 in the form of an Enterprise JavaBean Development Kit. As soon as the specifications for the APIs are final, a number of application and development tool vendors, such as Oracle, Sybase, BEA Systems, IBM, and Borland International, will begin exposing their tools and applications to Enterprise JavaBean APIs.

13.4.7   Other Flavors of Java

The Java chip is a CPU from Sun that executes Java byte code natively. It is based on Sun's picoJava architecture and is expected to be used in a wide range of devices, from small, hand-held applications to desktop network computers. MicroJava is a low-end implementation of Sun's Java chip. UltraJava is a high-end version designed for desktop machines. UltraJava is expected to have built-in MPEG compression.

Visual J++ is a Windows-based Java development system from Microsoft that supports Microsoft's ActiveX controls and includes a JIT compiler.

HotJava is a Web browser from Sun that is written entirely in Java and designed to run on JavaOS or a Java-based Network Computer.

13.5   Java and ActiveX

It seems like the debate is always ActiveX vs. Java, but in fact they are not really in the same category of tools. Java is a language. ActiveX is a component architecture for building applications. The JavaBeans specification is the analog to ActiveX, as it provides a component architecture for Java programmers.

Today these two products really do complement each other. For components that are used by many Web pages (such as dropdown lists and buttons), using ActiveX controls makes sense. The control is downloaded once, and the stored precompiled version is used as needed. For processes that require logic, and therefore code, Java is the right choice. For code that should not remain on a user's machine, Java is the only choice. It comes down to "Use the right technology for the job."

Netscape Communications has, to date, refused to provide native support for ActiveX controls. Netscape cites as its reason that ActiveX technology puts a proprietary framework on the Internet. However, Canadian toolmaker Ncompass Labs has developed an ActiveX plug-in for Netscape.

ActiveX will be the easiest migration for many Windows shops because it integrates with Microsoft desktop application tools; Java will be the choice for those requiring portability.

However, both Java and ActiveX will flourish as i-nets proliferate. In fact, many companies may experiment with both technologies as they mature and attract more third-party support.

14

I-Net Management

Corporations continue to deploy intranets at an unprecedented rate because of the technology's scalability and utility in information sharing. However, IT departments usually are inadequately trained in the actual management of an intranet.

Tactical solutions, not long-term strategy, are what companies should focus on in managing their intranets—and the sooner in the intranet development process, the better. Management by troubleshooting doesn't work in this environment, nor does long-term planning (technology is changing too fast). Instead, companies must work on integrating the different management disciplines to manage bandwidth, traffic, security, storage, and content—the five components of i-net management—to achieve the desired benefits.

For example, traditional system management focuses on network performance, application performance, and response time. In intranet environments, the focus should be on bandwidth and traffic—most organizations just don't have enough bandwidth to handle the traffic the intranet is going to generate—with an ever-watchful eye on security.

In many organizations, either intranets popped up in grassroots fashion around the company or the intranet was already up and running before management issues were recognized. Today, intranet users tend to be either on the bleeding edge, grappling with managing their sophisticated intranets, or just beginning to address intranet development and management as they hear reports of success. IT organizations need to take control and scale back on new features while management concerns are addressed.

Data backup becomes a major issue for organizations. Web sites use large image files and sound files. These are often static files, and the efficiency of the network can often be improved by storing them on a secondary storage site. Many organizations are turning to hierarchical storage management systems to reduce the overall storage requirements of the Web. Hierarchical storage management systems automatically move files from hard disk to slower, less expensive storage media. The typical hierarchy is from magnetic disk to optical disk to tape. The software constantly monitors hard disk capacity and moves data from one storage level to the next based on age, category, and other criteria as specified by the network or system administrator. When a file is moved off the hard disk, it is replaced with a small stub file that indicates where the original file is located.

As more and more Web sites are used for distribution, selling, and providing customer service, availability becomes more critical. Once a revenue stream or competitive advantage becomes Web-server-dependent, that server has to be accessible.

But security continues to be the number one issue for i-net management. A Net server could be hit (visited) 100,000 times a day. For each hit, the organization wants to be assured that the Web server can be accessed and that the data and Web pages accessed are correct.

IT managers should create clear policies concerning the legal responsibility and intended use of an intranet and keep a close inventory of all intranet hardware and software components. As many intranet management functions as possible should be automated.

The plus side of having all these additional issues to deal with is that the Web actually makes it easier to deal with them. The Web platform can be used to distribute management information so that it is accessible remotely from any console, not just from a few dedicated ones in some locked room near the computer hardware. System administrators can check on the health of the system from anywhere at any time. They can spot troubles before users are affected by them. Support staff can check the same data as calls come in so that users can be given information when they call in—no more "I will have to get back to you" from the support staff.

14.1 Network Management

It seems that the job of the network manager has come full circle. In the days of the glass windows, computers were centralized and easily controlled by the MIS department (the EDP department). Information was not shared among users, it was shared among applications and controlled by the applications themselves.

With the advent of Ethernet, local area networks sprang up all over, creating the reverse situation: Users could easily share and access information across the organization, but network managers were faced with the difficult task of controlling costs and ensuring network uptime. Network managers have begun to recapture control of their organizations' servers, but, with the explosion of corporate intranets, they are about to lose control again.

As the push toward the Internet, intranets, and the Web continues full speed ahead, network managers are confronting issues such as how to authenticate not only users, but also Java applets and ActiveX components that come with Web pages, and how to integrate Web front ends into all existing mail, file, print, and database applications. Users are expecting Web-enabled services on the network. Most vendors with Web-enabled service-oriented products are releasing back-end versions of their products for more and more platforms.

As use of the Web increased, network management software vendors such as 3Com Corp., Bay Networks, Inc., Cisco Systems, Inc., and Cabletron Systems, Inc., adapted their network management software to allow reports to be delivered via the Web. This allowed network administrators to share status and performance information with the people who needed it, regardless of where they were—or what platform they were on.

14.2 Bandwidth Management

In addition, intranets seem to have a life of their own. Users with Netscape on their desktops and with easy-to-use software build HTML (Hypertext Markup Language) pages. They pass the URLs (uniform resource locators) to colleagues so that everyone can share the information. Soon there are so many users accessing the Web pages that traffic can come to a virtual standstill on that particular segment of the LAN.

And this is only the beginning. Technologies such as "publish and subscribe" and push are really going to affect network traffic.

The intranet phenomenon can be managed through bandwidth

distribution and bandwidth management, which means dealing with network management of a virtual environment at the application layer, not the traditional device layer.

Bandwidth distribution involves moving more bandwidth to the edge of the network. It means using switching and higher-speed LAN technologies. It means implementing switched Ethernet—or shared or switched Fast Ethernet—to the desktop. This, in turn, requires a strategy to upgrade corporate backbones and perhaps WAN links. It also requires installation of the necessary hooks in various network elements to support bandwidth management.

Bandwidth management involves stratifying users according to who will get network access priority in the bandwidth-finite world—setting priorities based on traffic type. It also includes classifying information to determine who gets access to what data. This is especially critical for applications such as videoconferencing that demand a high quality of service. Establishing policies for information access can better ensure network reliability and performance.

14.3   System Management

The development of system management software for this new platform has been slow. Since most early Web implementations dealt with static information, it wasn't cost-effective to worry about managing the system. As organizations begin to use their i-nets to let customers place orders that then get processed by the back-end systems, system management is becoming a big issue. These systems need to be managed—they are affecting inventory, extension of credit, and committing to shipping dates.

System management applications collect data from distributed resources and then initiate actions based on this data. Hardware, operating systems, databases, and even management applications have their own schemes for discovering and recording their pertinent data.

14.3.1   More Than Just Collecting Data

Collecting data is not enough. To be effective, the collected data needs to be turned into information, acted upon automatically, and shared among applications—all under one umbrella.

The problem of collection—and knowing what data to collect as well as from where—is compounded by the Web. Servers get overloaded, traffic gets bottlenecked; users come in via remote access, via internal access, and through the firewall from the Internet. The variables increase exponentially.

Most organizations were looking for tools to help them manage performance—CPU usage and disk I/O rates. Now organizations are looking for products that are based on Internet technology and can handle all the necessary functions: Monitor systems-level functions; monitor Web-server functions; support a variety of operating environments; integrate alert systems from disparate sources; support remote access for system management; and support pagers as an alert alternative.

Organizations have just begun to use systems management standards for managing the rest of their IT environment. Those standards and products that support those standards have been long in coming—and implementation is not without its own perils. Systems management software has never been a major focus for hardware vendors because they make their money selling servers and network components. Software vendors that sell such products have an uphill battle convincing organizations about the cost benefits of systems management software.

Often the level of integration offered by the software is limited to the ability to launch various monitoring applications from a single screen or to display various types of alarms on a single map. Integration should result in the sharing of intelligence between applications.

What kind of intelligence? In a typical networked database application, all the network elements that sit between the user and the application must be in good working order for a user to get a quick response time from the application. Today's management platforms are capable of alerting network managers if any link in that chain fails.

But suppose no single element in the chain fails, and yet multiple elements are beginning to show subcritical degradation in performance. The cumulative effect is that the response time to the user drops precipitously. Yet the network manager probably won't detect a problem unless a user calls to complain. That's because the management console is unable to logically link the reduced performance of multiple elements into a consolidated view of a particular network service.

Blame for failure to provide this type of interelement logic can't all be laid at the feet of the platform vendors. Each vendor has tried to provide its own application processing interfaces (APIs) and data models to enable a more integrated framework for managing heterogeneous networks. But the hardware vendors (who write the applications that monitor their own products) and third-party management software vendors have needed standards to solve the problem.

Everyone seems to be looking to the Web to provide solutions to problems. Enterprise management is no exception. Organizations are looking for Web-based tools to monitor and distribute management data. Among the many characteristics of Web-based management are platform independence, a common GUI (graphical user interface), and the ability for technicians to check on devices when they are away from their desks. But Web-based management provides a platform—it doesn't provide integration. The two major Web protocols—the HTML protocol that acts as the standard for creating Web pages, and HTTP (Hypertext Transfer Protocol), which delivers data from sources to destinations in the Web architecture—also have basic shortcomings when it comes to real-time management.

HTTP is a stateless protocol. A new TCP connection needs to be set up each time the management software wants to get data. In addition, it makes asynchronous communication such as SNMP traps impossible.

Companies with their feet firmly planted in this technology are trying to rectify the situation. Sun Microsystems and others have proposed Java Management Application Programming Interface (JMAPI), a set of extensible objects and methods for the writing of Java programs for managing over intranets. Microsoft Corp. and others have proposed Web-based Enterprise Management.

14.3.2   SNMP and DMI

Simple Network Management Protocol and Desktop Management Interface were developed to standardize the way all printers, PCs, routers, and platforms report on themselves. They have become the accepted nonproprietary means of collecting raw data from resources and sharing that data among platforms. SNMP, developed by the Internet Engineering Task Force (IETF), has become the common denominator for sharing data among management products, and DMI, developed by the Desktop Management Task Force (DMTF), has become the standard for reporting PC data. DMI compliance, however, provides for interoperability rather than true application-to-application data integration. It ensures that PCs and desktop software from different vendors will be able to report on themselves to asset management applications from different vendors.

SNMP, which is widely used to gather information from network devices, has been around the longest. It is based on TCP/IP standards and provides a common format for network devices to use to exchange management information with the network-management station (or stations).

But it certainly wasn't developed with the current architectures in mind. The station controls the network through agents located in the different pieces of equipment on the network. The management software polls each SNMP agent module installed on each network component for status information. This data is stored in the management information base (MIB), which defines the structure of the collection of managed components (objects). Just consider the network traffic that generates!

DMI provides a different data model from SNMP's. This difference tends to keep desktop and PC server information isolated from the bulk of network health statistics collected and displayed on enterprise management consoles. And SNMP does nothing to help integrate or correlate data collected from various components in the enterprise environment.

SNMP is static and contains no imperatives for action. It only gets information and sets variables in the information. The format of its data must be incorporated into management applications that expect to use this data. The model presumes that a management application will retrieve the data and use it to notify humans of events that require action.

DMI is slightly less passive than SNMP but has no scheduling or grouping of operations. It is designed more for desktops than for enterprise automation of management functions.

14.3.3   Web-based Enterprise Management Specifications

Seventy vendors, including Intel, Cisco Systems, Compaq, and Microsoft, have formed the Web-based Enterprise Management (WBEM) group to integrate Web perspectives into management standards. The WBEM group proposes to integrate the two prevailing management standards— SNMP, which covers networking equipment and software, and DMI, which focuses on desktops and servers—with the Web's HTTP to create a new Web-enabled and Web-based enterprise management architecture.

Although SNMP is a standard protocol, vendors provide extensions to the information reported by their equipment. Because of this, there is such disparity among the network-management platforms that integration across platforms is difficult. In addition, application developers have to develop a specific version for each platform, raising costs and development time.

The WBEM group seeks to ensure the availability of a standard architecture that provides better integration between applications and protocols, cheaper tools, and more innovation. Management applications today cannot support associations or relationships among data from unexpected sources because they were not configured to understand this data. WBEM has self-describing data that includes the relationships among devices and

Figure 14-1
Architecture of Web-
based Enterprise
Management

Web management console					
HMMP					
DMI		SNMP		HMMS	
Applications	Machine	Network card	Hub port	Hub card	Hub
Managed objects					

other resources. This means that WBEM applications will accept and use data from recently installed resources and from new types of resources.

WBEM isn't a product, a standard, or even a detailed technology. What WBEM offers is a standard approach for a software structure that would run on any platform and deliver device-level services to SNMP platforms.

WBEM's protocol, illustrated in Fig. 14-1, provides a common means for applications to execute remote tasks and be certain of hearing the results. Today's management applications can't command action and have to rely on a compatible remote agent to decipher and carry out the command.

Today, each managing application speaks its own language, understood only by its agents, which carry out their useful management tasks. Foreign agents (those built by another vendor) get minimal attention. WBEM replaces this with a common language that is richer than SNMP and designed for sharing data and executing management tasks. HTML and Web browsers provide a common means to display information.

Management applications today require a mechanism to transport and collect data. Without transport, the raw device data just sits on the device. WBEM offers the Web as the transport mechanism, supplanting or coexisting with the proprietary framework or management platform in use today.

The WBEM group is proposing a Hypermedia Management Architecture, currently based on three Web-based standards:

■ *Hypermedia Management Protocol* (*HMMP*) provides network- and systems-management data to browsers; it is being reviewed by the Internet Engineering Task Force. HMMP lets browsers access and receive systems- and network-management data from devices and applications.

■ *Hypermedia Management Schema* (*HMMS*) is a data model used to represent managed objects via the Internet; the Desktop Management Task Force expects to include this in its next version of DMI. HMMS is now known as the common information model (CIM). This specification and products using it are likely to appear first.

- *Hypermedia Object Manager* (*HMOM*) gathers management data from applications to be displayed on central management consoles.

The WBEM protocol, schema, and object model specifications allow management hardware and software to share data. These specifications can be viewed at www.wbem.freerange.com.

Common Information Model CIM, previously known as Hypermedia Management Schema, defines a common data model that represents the managed environment. Essentially, it defines a structure for device home pages, describes the elements of the computing environment, and records the associations and relationships among the elements. CIM objects are self-defining. An application does not have to be preconfigured for the type of data it will receive from a device or other managed object. This lets applications deal with unexpected new types of objects.

CIM employs basic object-oriented constructs. The core schema establishes a basic classification of the elements of the managed environment. The four core classes are system elements, applications, resources, and network components. The managed elements can be physical, logical, or systems.

By treating the physical components, logical constructs, and associations as objects, CIM provides hierarchy and inheritance and promotes flexibility because the classes can be extended. The model can be extended through class extensions, by adding properties, by overriding, and with polymorphism.

Microsoft claims its Common Information Model Object Manager will ship with both Windows NT 5.0 and Windows 98, both due out in 1998.

Hypermedia Management Protocol The design of HMMP is based on HTTP. HMMP collects and transports the raw data that populates the database. It runs over the ubiquitous HTTP, and therefore does not require a management platform to provide transport and collection services. Whereas SNMP relies on passive gets and puts, HMMP has imperative operations designed to let management applications take action, such as executing a method or applying a filter. All of the operations can target single objects or collections of objects that are associated by geography, type, or some other category. For example, the filter will find all objects having association with or reference to an object, and then the method_exec will execute a remote task for the group.

The HMMP event architecture allows for acknowledged delivery, guaranteed delivery, event correlation, and scheduling of management actions—functions not provided by SNMP or DMI. With SNMP or DMI, management applications that require delivery of messages or that need to schedule management tasks in reaction to events have to rely today on manager-agent conversations.

With the move to CIM and the name shortening of HMOM, it is likely that HMMP will be renamed before it is standardized.

Hypermedia Object Manager   HMOM, now known simply as Object Manager (OM), takes the CIM data model, which is a definition, and turns it into a database that applications can use. Applications subscribe to the OM for access to the database. OM manages elements as objects, and integrates management data and coordinates control through a variety of management protocols and interfaces. Information is presented through a Web browser or to HMMP-enabled management applications.

Hypermedia Management Architecture   Here's how this management architecture works. Using a Hypermedia Management Application, or HMMA, a Web browser links directly to a device or application, called a Hypermedia Managed Object (HMMO), via HTTP. Each HMMO has its own URL. An HMMO provides management information in an HTML Web page. The underlying management protocols on an HMMO, such as SNMP and DMI, would not matter to the network manager. As long as there is an HTML interface on the device or application, the information can be accessed. And the HMMA can access non-HMMO devices that just run SNMP or DMI.

Put another way, each managed device has its own Web page. A browser can access any device via HTTP, and from there examine the data.

Three Models of HMMA   WBEM has also presented three basic models for using the Hypermedia Management Architecture:

- *Simple Browser-based Management.* Devices structure the management information according to the HMMS data model and deliver the data directly to a browser that has been designed for that particular HMMS data model. The browser will present each device in its own Web page.
- *Advanced Browser-based Management.* This approach inserts HMOM between the devices and the browser. HMOM can accept data over protocols such as SNMP and DMI, associate the data using the HMMS model of the environment, and forward the data to a browser via HTTP.
- *HMMP-based Management Applications.* In this case, rather than forwarding management information to a browser, HMOM forwards information to a management application. This application might signal

corrective action to be taken on a device or system. HMMP signals the corrective action and returns the results to the management application.

BMC Software intends to deliver the first WBEM product, as a component of its PatrolWatch, a Web-browser-based console. PatrolWatch is an example of the Advanced Browser-based Management model. Patrol products use Knowledge Modules to collect and analyze raw device data. In PatrolWatch, a OM management services layer which formats and delivers the information to the PatrolWatch console has been included. BMC will extend the next PatrolWatch release to fully use the CIM but is tying delivery of the code to the release of NT 5.0.

14.3.4 Java Management Application Programming Interface

Corporate networks are beginning to shift to Web-based management strategies, which allow remote access and give multiple users access to management data. But for Web-based management strategies to become more prevalent, a unified approach to systems, applications, and networks is needed. WBEM defines three interfaces that create a common data structure—an object-oriented representation of applications and communications protocol. But WBEM fails to address the question of where network management statistics are generated. The Java Management API (JMAPI) from Sun Microsystems aims to help developers solve this dilemma by creating applets that can troubleshoot systems, applications, and devices. The purpose of the API is to enable network managers to monitor systems and the network in real time via the Internet.

JMAPI integrates applications and provides network-management capabilities to enable businesses to "manage the Web with the Web." JMAPI, a specification for developing Java-based management applications, has already been adopted by more than 30 industry leaders. JMAPI specifies a browser-based application integration framework for a common look, feel, and application behavior. JMAPI-based solutions (written in Java) support data sharing and result in faster application development. The use of Java applets as monitoring agents is attractive from an architectural point of view, since it off-loads processing and presentation burdens from the management console's CPU.

The Java Management API provides objects and methods for managing enterprise networks over the Internet. JMAPI is a set of interfaces that read and write Java objects, such as status information, directly into a relational database for management. The APIs use remote method invocation as their remote communication mechanism. JMAPI will have hooks so that users can run different pieces of applications together.

The Java Management API provides many of the association and interoperability benefits provided by WBEM, but Java platforms are currently regarded as a more open environment than the Microsoft and Unix environments.

JMAPI supports SNMP using an SNMP interface built into the API that allows objects to contain information they get from SNMP agents. There currently is no support for DMI.

Like WBEM, JMAPI promises location and platform independence. The technology, according to Sun, provides for "write-once, run anywhere" software. In the Java architecture, software changes and upgrades are dynamically downloaded to the client when the client requests access to an applet. This is in contrast to the problems that occur when there are new versions of protocols such as SNMP or HMMP. In other words, Sun is saying that there is no need for HMMP in JMAPI.

SunSoft's Solstice Workshop is a developer toolkit that provides a set of Java-powered objects and a methodology for creating Java applets that can manage an enterprise network over intranets and the Internet. The components of Workshop are the Java Management API, a simple ODBC-compliant database, and the Java programming environment. The APIs allow developers to write systems- and network-management applications or applets. The applets, which can run on any Java-enabled operating system, collect information about managed devices. The information is then available to any Java-enabled browser. The classes and interfaces that make up Java Management API are already in use by other vendor development teams and were released in beta in late 1997.

14.3.5 Interim Solutions

The competition between Microsoft (WBEM) and Sun (JMAPI) in the Web-technology market could slow the standards process. Each is addressing a different set of issues. However, JMAPI and HMMS are not mutually exclusive. This may be a situation in which the end user will see a hybrid approach. Java is a protocol-agnostic programming language, whereas WBEM is primarily concerned with a management protocol that works with Web browsers.

While the battle of the standards goes on, existing systems-management software is beginning to offer Internet capabilities. Tivoli offers Tivoli/net.Commander, which allows users to directly manage their intranet applications as well as the underlying Internet platform of Web, news, mail, and other services. Computer Associates has enhanced its CA-Unicenter systems-management product with the ability to manage Web servers for the intranet or Internet. Compuware Corp. has added Internet and intranet server management features to its EcoTools line of network applications management tools.

WireTap, a performance-management tool from Platinum Technology, can be used to monitor traffic to and from corporate intranets and Internet and Web servers. Like a network sniffer, the tool examines network utilization by network protocol, application, and other traffic categories, and provides response time and occurrence statistics for service requests, such as requests for intranet and Web pages.

BMC Software Inc.'s Patrol has an Internet Knowledge Module that can be used to monitor electronic-mail servers and gateways. All modules of the product, including those that monitor file capacity, can be viewed from one screen. Although BMC Software's Web-based remote product, PatrolWatch for Web Browsers, is in beta testing, few management console applications offer remote notification in a modularized fashion out of the box. Like most point solutions, they are add-ons that often need to be customized and are short-lived.

Users and vendors say that the natural progression of systems management through the intranet will be to address more network concerns, such as bandwidth and traffic management. The things that are monitored change. The perimeters of the network change with a variety of access methods. Load balancing between servers becomes a bigger issue—no one server should be a single point of failure. Access speed between client and server needs to be monitored.

System administrators want to run management software via a Web browser from anywhere and have it be a single point of administration. Administrators don't think they are asking for much—just the ability to manage the network remotely using a Web browser. And they admit they can do this using piecemeal solutions. But what they want is integrated systems- and network-management solutions. And they are hoping that they get their wish in 1998.

Remote Maintenance   With machines and software spread throughout an organization, maintenance—upgrading software and virus scans, for example—is a nightmare. The IT organization can physically visit each machine and perform the work or do it remotely from one central site. Remote maintenance is one of the advantages of network computers (see Sec. 9.1.1).

Remote maintenance is becoming an option with PCs as well, with new adapter cards as well as agent software from Intel Corp. The new Ether-Express card lets managers remotely turn on, boot up, and service PCs from a central location. The card uses IBM's Wake-on-LAN technology, which is also being used in network computers.

Intel also offers LANDesk Service Agent, which is a preboot agent that enables new PCs to be booted from a remote management server for installation or upgrading of software (operating systems as well as applications).

14.4   Document Management

As more companies build intranets, network administrators are discovering that deploying collaboration applications has led to a new management problem: tracking and controlling the thousands of documents that are proliferating across internal World Wide Web sites.

The bulk of the information on intranets is made up of documents, from the phone directory to the cafeteria lunch schedule to policies and procedures manuals to design specifications. Version control, access privileges, and reconciling the changes of multiple authors are just a few of the issues that arise when documents are shared. If these issues are ignored, many of the benefits of collaboration and intranets may never be realized.

Organizations turned to document-management software to create and maintain the formatted documents that were distributed to an entire organization. This tended to be proprietary software, and pages created for one document manager often couldn't be viewed on another. Applications very quickly came to have a departmental focus rather than an enterprise focus.

Intranets jump past these technicalities, making nearly every piece of information on a network available corporation-wide and using a browser that makes everything look "equal." In addition, since documents can be put into an intranet so easily, system managers suddenly need to address access, control, and authorship issues.

As organizations begin to implement document-based intranets, most of the documents are managed by the department that created them. It is difficult to keep every copy within the network current because users must remember to check whether there is a later version.

One step in the control process is to have one group responsible for all the documents published. But this can quickly become a nightmare without effective document management tools. Organizations are turning to software that will automate document management, support project collaboration, and offer search and index functions.

The software should have an automated method for managing versions and user access to the documents. It must monitor not only the users who had access to certain documents but also the level of clearance they have with those documents. One group might have read-only access to some documents, for instance.

Companies such as Open Text Corp., SoftQuad International Inc., and Adra Systems Inc. are outfitting their document-management tools with browser interfaces. Lotus Development Corp. has even added a document-management feature to Domino. Domino.Doc will add document checkin/checkout and version control to Domino, making documents accessible using Web browsers or Notes clients. Domino.Doc enables IT departments to develop document-management applications quickly on existing Notes networks.

Other vendors are sure to follow. According to International Data Corp. in Framingham, Massachusetts, Web-enabled document-management applications jumped from zero users in 1995 to almost 94,000 in 1996.

Document management is not usually a consideration in the beginning. It doesn't seem like a necessity until managers realize how hard it can be to keep track of changes in this unstructured environment.

14.4.1 Content Management vs. Document Management

Don't confuse document- and content-management products. Document-management tools are used to create and keep track of the documents that are distributed on an intranet, but don't do much to manage the content of the World Wide Web. They focus on enterprise knowledge management.

Content-management tools, which have begun to emerge over the past year, perform similarly to document-management tools but are limited to the intranet, meaning that they do not factor in documents that reside in legacy or Unix systems. These tools will also place more of an emphasis on the management of intranet-specific information, such as the validating of hypertext links. They focus on Web-site information.

However, the line between content and document management is blurring, and the distinction may soon disappear altogether. In many corporations, processes that once resulted in the production of a document now culminate in a Web page instead, making it critical that IT managers identify the right tools for the project.

14.5   Quality of Service

Users need to somehow reserve network bandwidth for high-speed traffic streams from videoconferencing or audio applications that can't tolerate delays. To support the different acceptable delays and bandwidth requirements of these applications, networks can use quality of service (QoS) to accept an application's network traffic and prioritize it. To accomplish this, the network device (such as a router or switch) requires a profile for the network connection that includes the minimum and maximum bandwidth required, the maximum acceptable delay, and, often, a relative priority.

Some of the characteristics of QoS include

- Reliable—there are no lost packets.
- Guaranteed—messages will be delivered immediately or eventually.
- Assured—a message will be delivered only once.
- Transactional—integration of messages that make up a unit of work (transaction) is supported.
- Priority—different types of messages can be granted different priorities.
- Reporting—reports of activities to determine QoS levels are provided.

Most networks currently offer little in the way of QoS support. Some router-based networks offer prioritized queues for traffic, but those queues cannot be parceled out to different users dynamically, nor do they offer sufficient granularity to support a variety of user requests. The Resource Reservation Protocol (RSVP) may be the answer.

14.5.1   Resource Reservation Protocol

RSVP is a new protocol for the Internet that has been sanctioned by the IETF because audio and traffic on the Internet are expected to increase.

For a premium price, RSVP will enable certain traffic, such as videoconferences, to be delivered before other traffic, such as E-mail, when the network gets congested. Today, all traffic moves on a first-come-first-served basis and is charged for at a flat rate. RSVP will begin to permit different qualities of service and differential prices. Figure 14-2 illustrates how RSVP might link an IP network to an ATM (asynchronous transfer mode) network.

Under RSVP, when an application needs guaranteed bandwidth or quality of service, it can request a network device, such as a switch, to reserve the necessary resources. Each RSVP reservation requires state infor-

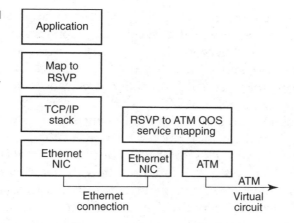

Figure 14-2
How RSVP Maps ATM
Services to IP
Addresses

mation in each router or switch that has reserved bandwidth for the reservation. Each router in the path between sender and receiver has to process the QoS request independently of other routers. Since no QoS-based routing protocols have been developed for IP nets, these independent routing decisions make it difficult to ensure that an optimal path will be assigned to data traffic. As more connections need more reservations, the network equipment gets overburdened maintaining all this state information.

This actually creates a dilemma: How are priorities set among the multiple reservation requests? RSVP doesn't create bandwidth; it allocates it. Given the cost of maintaining state information, is it possible to grant every reservation request? If users start employing agents to make reservations for them, it's easy to imagine agents attempting to do contradictory things. For RSVP to be truly effective, it will need a way to arbitrate between reservation requests.

The initial plans for RSVP specify four QoS classes—guaranteed, predictive, controlled load, and best effort—but only the first two classes have been included in IETF draft documents pending review and approval. (The last class, best effort, is simply the absence of RSVP service.)

RSVP has been quietly moving through the process of becoming an Internet standard, pushed by Cisco Systems Inc., which makes the routers that direct most Internet traffic, and by Intel, which wants to spur demand for microprocessors by making computers more useful for things like videoconferencing. RSVP is expected to be used by most Cisco customers by the end of 1998. However, before RSVP can be useful, the technology must be built into the operating systems of the computers that receive data via the Internet.

Market response to the IETF-approved RSVP has been slow. RSVP negotiates a quality of service level for each individual application information stream, which makes it particularly effective for applications that have long connections and very ineffective otherwise. Applications have to be classified as "special" or ordinary for the reservations to work. As the number of "special" applications grows (and they are all "special" with a high priority to their users, aren't they?), these applications compete for resources, and no one gets an acceptable response. RSVP handles the transmission from traffic source to destination and back. If anything along the way isn't RSVP-aware, it can't participate in the prioritization, which undermines the reservation itself.

Unless all major ISPs implement the protocol, "real time" is not possible. ISPs also have to be very cautious in how they deploy RVSP. They need to put limits on how much of their infrastructure customers can reserve, since the pipes are actually shared among their customers.

Many organizations are dodging the bandwidth scarcity issue by using high-capacity networking technologies such as ATM. ATM transmits cells that are larger than packets and better suited to simultaneously transmitting voice, data, and video traffic. And ATM products with quality of service guarantees are available now, but still for a high price.

Push Technology

We can't get around it—we're in the information age. Information is key to our businesses and our lives. We want lots of information, and we are sometimes fearful that we might miss the really important nugget, so we keep searching, not really sure when to stop.

For example, an analyst in charge of tracking competitors logs onto the Internet every day (at about the same time) and does a search using major competitors' names as the search string. And since each major search engine searches differ items or portions of items, the analyst checks using three or four search engines—just to be sure nothing is getting overlooked.

Push technology turns this process inside out. Users register what their areas of interest are (usually by subscribing to channels, which are categories of information), and the push service providers send daily feeds of everything that appears in those areas. The user gets an E-mail that information has arrived or can actually have the display incorporated into his or her desktop.

Organizations began seeing this as a way to distribute information internally as well. Instead of sending mass E-mails to everyone in the company, they push the information to each employee as a display. E-mails require employees to open the message, which they can postpone as long as they wish, creating a time delay, or they can even just delete a message without reading it. In order for an employee to get the information in the E-mail, the employee must be proactive—checking for messages and then reading them.

Internal push implementations take advantage of desktop technologies and allow employees to be reactive. The information is presented to them, as opposed to having them take actions to receive it.

For some, this becomes information overload. For others, it becomes an administrative task—what areas am I interested in? For still others, it seems like the perfect marriage between desktop technologies and information retrieval.

15

Push
Technology

I-net technology promises to make information available. Users search for needed information, always fearful that they will miss something of importance. Push technology—also known as Webcasting—promises to cut information overload by delivering the information a user needs. Push technology products—also called publish and subscribe—are designed to reduce the time spent searching on a network by automatically delivering information to users' desktops. Some users subscribe to third-party information providers that offer channels of information. Organizations can create their own channels to transmit only information deemed relevant to workers.

Some organizations have discovered that a pushed message prompts an employee to search for additional information, thus prompting more browsing.

15.1 Overview of Push Technology

The term *push* refers to an architecture in which a user receives information or applications automatically from a networked server. The term is used to describe a wide spectrum of products and services. How push happens and what is delivered vary among different push implementations.

To find information on a network (Internet, intranet, etc.), a user initiates a search by specifying a particular topic at an arbitrary starting point and at an arbitrary point in time. Multiple topics may require multiple searches. And since search engines check different information sources, users often repeat searches using different engines. The information that turns up may be dated or may not be pertinent, or some important information may not be captured because it is not reviewed by the search engine used or is considered old news and therefore ignored.

Push technology turns this traditional "pull" information-retrieval model around. Instead of the user's finding information, the information finds the user. Push services work off a user's information profile. They filter information on the basis of the subjects a user is interested in. Ideally, the push server delivers what is needed before the user even begins to look for it.

Traditional networks are passive; data is stored so that a user can retrieve it. Push architectures use the concept of *dynamic personal assistants* (agents) to access channels of information looking for information a user has indicated is of interest and to check Web pages of interest for changes. Only those that have changed are reported to the user.

Push technology is event-driven. Consumers are notified of each update without having to ask "What's new?"—it arrives at their desk automatically. This decreases bandwidth usage, as it is an asynchronous process. Users don't use up bandwidth searching and checking to see if there is anything new of interest. The publisher of the information just delivers it. It is like the newspaper waiting for you outside your door in the morning, but the only thing in this "paper" is the news that you care about.

There is a lot of anonymity in publish-and-subscribe technology. Publishers make the information available on a channel or tag. Subscribers receive information on a channel or tag. Publishers don't know who is accessing their material. Subscribers don't know where the information came from.

15.2 Types of Push Implementation

There are three types of push implementations:

- Blanket push
- Filtered push
- Publish-and-subscribe push

15.2.1 Blanket Push

Blanket push architecture is illustrated in Fig. 15-1. The client software is proprietary, and the server is the information publisher's server, which has gathered all the news stories to be published. The software issues periodic requests for updates from the server. All updates are pushed back to the client.

Many users turn to this push technology because the client software is typically free and it is a good way to get news feeds from a variety of sources. However, the constant need for Internet access fills the network bandwidth. These Web pages usually come with banners or advertisements, which are distracting and cause delays in page retrieval. With this type of implementation, users cannot be sent internal data via the push architecture, nor can the data be customized for the user.

15.2.2 Filtered Push

Filtered push architecture is illustrated in Fig. 15-2. In this case, the user specifies what the requirements are. The news wire stories get pushed once to the internal push server. From there, the data gets pushed to individual users based on their requirements.

This architecture allows an organization to customize external information with internal information. Since only one set of updates is sent

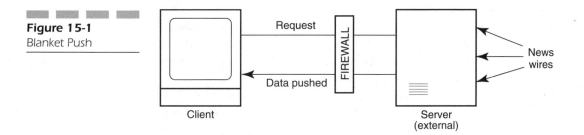

Figure 15-1
Blanket Push

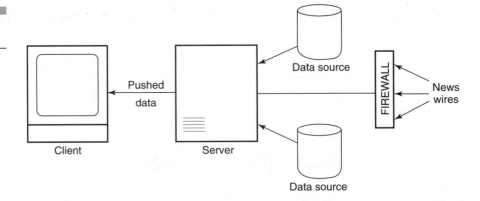

Figure 15-2
Filtered Push

into the organization (instead of there being an Internet connection for each user), network bandwidth isn't affected as much. However, this architecture requires more administration at the internal server, and there are no proven management tools to assist in that task.

15.2.3   Publish and Subscribe

Most organizations are trying to get to this model. As illustrated in Fig. 15-3, the information on the server is organized and contains both news wire stories and links to internal data. Users subscribe to a particular category (or two or three) of information.

This architecture puts the responsibility for managing the information on the push server onto the users and management. Tools to help with the management aren't mature yet.

15.3   Channels

A channel is a logical grouping (or category) of information offered by a content provider such as Reuters' financial news service. The notion of channels is also referred to as metacontent, or information about information. Web sites are very static; channels have a virtually continuous flow of information. Channel aggregators such as Incisa and PointCast collect content from various sources and serve the content to the push client (usually their own).

Channel content is often just a headline or a summary that points to a complete version of the article. In an ever-growing Web, channels provide ways to absorb information quickly and pinpoint useful content.

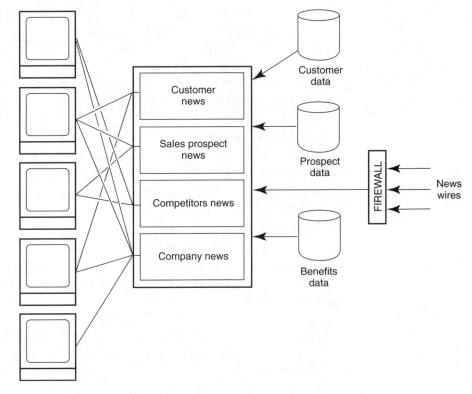

Figure 15-3
Publish and Subscribe

Most push products require a proprietary client and server component. However, server-based push technologies do not require any special client software. They deliver information via E-mail or standard Web browsers.

The inverse of the client-independent push model is the "push desktop" metaphor used in the Microsoft Internet Explorer 4.0 and Netscape Netcaster (see Sec. 16.2). These next-generation clients will provide a single interface to access multiple push sources.

15.4   How Push Works

The current push craze grew out of Internet-based content services such as those provided by PointCast Inc., which deliver various news and information channels in an easy-to-use interface. The user installs the software and indicates topics of interest, and then PointCast takes over, automatically updating the content on a scheduled basis.

The architecture for the concept of push has three major functions:

- *Accept incoming data.*   PointCast originally defined a user's notion of push; push clients have been equated with Internet-based services that deliver information via screen savers and stock tickers running along the bottom of the screen. In reality, information pushed to the desktop can come in all shapes and sizes, including HTML (Hypertext Markup Language) links, pop-up alerts, Java applets, and even Shockwave animation. From there the user clicks to follow links to news- or business-related notifications.

- *Access the channel.*   Data comes to a push client via specific channels. Users pick and choose which channel their push client is tuned to. Administrators build channels for intranet use and are responsible for those channels' content.

- *Push information to the users.*   Most push servers actually use a pull technique: The client software periodically sends out a request for updated information. In most cases, systems using client polling have a minimal impact on network performance as long as clients are connected to a local proxy server. In contrast, Incisa uses a multicast approach to send messages to clients. But functionally users don't see a difference between the two approaches unless information really is needed instantaneously, such as in a real-time financial services application.

PointCast Network broadcasts news stories—business updates, stock quotes, news headlines, etc.—to any user who has PointCast's client software. PointCast channels are category-based—finance, sports, etc. Any changes within the Web pages of the selected category are sent to the user's desktop. If a user chooses sports, any changes to sports-related Web pages will be delivered—even if the only sport the user is really interested in is soccer.

The PointCast software, which is downloadable for free from Point-Cast's Internet site (www.pointcast.com), includes a screen saver that flashes the latest items on the monitor when the computer is idle. A click on the headline takes the user to the article, which has been downloaded to the PC. The software can be given limits on how much space it can use and will overwrite old stories when it needs more space.

Today there are more than a dozen push clients for receiving general-interest information over the Internet in scrolling tickers, screen savers, pop-up messages, or Web pages.

Push servers actively distribute information (some can also distribute applications) that matches a user's interest profile. In contrast, Web servers and file servers passively wait to be asked to find what the user is looking for.

Microsoft Internet Explorer 4.0 and Netscape's Internet client called Netcaster both incorporate push features that enhance the Web browser as an information-harvesting tool. Microsoft uses its proposed standard for push—the Channel Definition Format (CDF). CDF will allow Web publishers to filter content and use third-party push products to do the actual delivery. Netscape currently doesn't feel that a separate push standard is necessary and uses HTML, Java, and JavaScript to deliver content in Netcaster.

15.4.1 Categories of Push Software

There is push client software and push server software. These categories are intended to be used on a corporate network. In some cases, push clients can work without a push server; the software works off the network itself. Push server software downloads the desired information to the push server. In this case, the push client gets its information from the push server.

Content providers offer services that broadcast general-interest content (channels) over the Internet. Most of these content services allow users to download client software and subscribe to information sources free. These providers usually use advertising to pay for these services.

Some companies, such as BackWeb and PointCast, provide push client and server products as well as offer content services.

Browsing software from Microsoft and Netscape uses push technology. As push technology begins to appear in the desktop suites, the definition of push client will become broader.

15.4.2 Importance of Directory Services

Directory servers can function as the containers for push-management instructions.

Administrators can use directories to store such information as

- Authority to subscribe to a channel
- Permission to self-subscribe to a set number of channels
- Authority to set up other people to subscribe to channels
- Authority to create channels
- Issuance of digital certificates, giving access to secure channels, to employees and external customers or partners

15.5 Push Standards

Microsoft's Channel Definition Format defines push content and how users receive that content. CDF facilitates an easy, standard way for Web sites to broadcast information to users.

CDF is a simple text-based format that allows publishers to define channels that point to Web pages—to push or Webcast content (information) to browsers. A CDF specification has three main components: the channel itself, items, and schedules. The text-based CDF file structure is based on Extensible Markup Language (XML), a subset of Standard Generalized Markup Language (SGML). XML is a lightweight language for structuring and describing data, whether for content or for applications. XML could offer cross-platform data exchange between software such as calendaring programs, contact databases, and browser bookmarks—something desperately needed in the mobile computing market. XML may also help standardize data formats for exchanging structured documents across extranets. In addition, the tags embedded in XML could provide the Web with a form of distributed metadata.

Many are betting that this technology will move the Web beyond the limits of static HTML. The specifications for XML were released in late 1997 by a working group of the World Wide Web Consortium, which is responsible for setting standards for the Web. Microsoft has based its CDF on XML, and Netscape has submitted a proposal to combine XML with Meta Content Framework, a Netscape-licensed Apple technology that describes Web content.

A complete description of the CDF proposal is available at www.microsoft.com/standards/cdf.htm.

CDF files describe what content is in a channel, what has changed, and when the client should check for updated information. CDF depends on content that already exists on a Web server; the channel file includes the unform resource locator (URL) that tells a CDF client where to check for newly published information. Though a channel defines a suggested time interval at which to check for updates, a CDF client can check the channel at any user-defined increment. Technically this is a pull rather than a push process—the client goes to the server to request any updates—but for the end user the experience is the same: Content comes to the user rather than the user searching for it.

To set up a CDF-based channel, a Web publisher creates a .CDF file that defines which Web pages are part of the channel. Web clients download the file via HTTP (Hypertext Transfer Protocol).

CDF is not a glorious new technology; it just provides a standard way to use channels. All push vendors have some method of defining, delivering, and displaying channels. CDF only defines a simple framework for push content delivery.

Microsoft has submitted CDF to the World Wide Web Consortium for consideration as an Internet standard. Major push technology vendors such as PointCast and BackWeb have backed CDF. Several Internet access providers have also lined up behind the CDF file specification, including America Online and CompuServe.

Not to be left out, Netscape has announced its Mercury and Apollo software, which will allow the direct delivery of Web content to the desktops of intranet and extranet users. And Netscape's Constellation technology is designed to allow users to customize the desktop delivery of Web content via user interface features such as rolling tickers or news headlines.

15.6   Push Client Technology

Push client software lets Internet and intranet users customize delivery of information directly to their desktops from a variety of sources. Push client software lets a user personalize information of interest, provides some level of real-time notification of updated content, and, in some cases, customizes the intranet infrastructure. Users can be alerted to changes in stock prices, for example, in real time.

Push clients as well as any Web browser can offer news and information. In fact, push isn't an accurate description of how push client programs operate. Push suggests that clients are notified automatically whenever new content is available. Currently, only Incisa, from Wayfarer Communications, comes close to that model because it uses a proprietary protocol to identify new content. Most push clients use a subscription model: Clients poll the content servers at predefined intervals to see if new content is available.

Push comes from server push, a term used to describe the streaming of updated Web-page contents from the Web server to the browser. This streaming is accomplished by keeping the connection between Web browser and server open after a Web page is downloaded.

Push clients run in background and automatically download information to the user's desktop. All are designed to deliver news and information; some can also deliver software or multimedia. Users customize the software to deliver only certain channels. When new information is delivered, push clients notify users through E-mail, playing a sound, displaying an icon on the desktop, popping up the application, or displaying headlines on a screen saver, wallpaper, or a scrolling ticker tape.

While the graphical user interfaces (GUIs) may look similar, the behind-the-scenes features are somewhat different and definitely proprietary. Some products use HTTP to download content index files; some also intercept HTTP requests between the browser and the TCP/IP stack. PointCast does not use a "linked" connection to the Web. Instead, hot links within PointCast material bring up a browser and take users to the appropriate Web page. Since Castanet typically sends data in HTML format, users can view their channel content through a standard Web browser.

However, push clients are only one piece of the equation. Each push client has a proprietary method for implementing content channels and usually makes a developers' kit available for that purpose. In some cases, content providers need to use special server software to offer a channel. For corporate intranets, where many desktops run push clients, some push client vendors offer a caching server (referred to as a push server) that runs locally on the intranet and provides shared content feeds for all the desktops, alleviating firewall traffic.

15.7   Push Server Architecture

Push servers allow an organization to set up and broadcast its own information channel in addition to receiving and broadcasting Internet channels. The advantage of using push technology to distribute information over an intranet is that the appropriate information is delivered immediately to people who can use it. However, these push servers need to be managed, and someone has to create content for them, which includes determining what external information is relevant for a user's job functions.

Push servers in an internal corporate network operate in the following ways:

- *Internal content creation.*   Internal push content is grouped together in individual channels. The administrator responsible for creating channel content generates the necessary files or information and submits the content to the server to be broadcast. The content can range from a simple text message to a complete Java application.

- *Subscription.*   From a centralized directory, a network user subscribes to a channel of interest. Some push servers allow the administrator to password-protect channels to make them accessible only to specified individuals. The network user must have a compatible push client installed on the local machine.

- *Notification.*   The user receives updated content in one of two ways: The push client software can automatically poll the server at a specified interval, or the server can send out the information immediately and the push client software—always "listening"—will receive the broadcast.

■ *External content services.* Most push products supply preconfigured channels of general-interest news and information. To access this content, the push server caches the channel information from the Internet locally, acting as a proxy server so that network clients don't need to access the Internet for every update request.

As organizations begin to take advantage of Internet-based technologies for their internal corporate networks, push technology will find a place in those intranets. Internal push servers, such as BackWeb, Marimba's Castanet, Wayfarer's Incisa, and PointCast I-Server, require their own proprietary clients. To reduce the amount of traffic over the wide area link to the Internet, these push servers include a proxy server feature or option to cache data from the Internet locally.

15.7.1   Unicast and Multicast

Most push applications, including PointCast, are unicast architectures. Each subscriber makes a separate IP connection to the provider's host, and each subscriber's session creates a separate data stream, even though the content of that stream may exactly duplicate another subscriber's session on the same subnet.

Unlike IP broadcast, which sends an identical data stream to all hosts on a network, IP multicast employs a one-to-many architecture instead. A host that wishes to receive an IP multicast session sends an Internet Group Management Protocol (IGMP) request to join a session, and is temporarily assigned the same IP address shared by all other subscribers to that session. Hosts that request the session actually receive it, and all subscribers to a given session receive a single, shared data stream. The advantage is bandwidth savings. And as the number of "shared" users grows, the savings increase.

TCP is a unicast architecture, so IP multicast uses User Datagram Protocol (UDP) as its transport protocol. Work is being done to develop native multicast transport protocols, such as Realtime Transport Protocol and Realtime Streaming Transport Protocol, which incorporate the data integrity and packet sequencing capabilities that are missing in UDP.

Organizations are turning to multicasting as a way to distribute information within a widely geographically dispersed employee base. Microsoft uses multicasting to allow employees to sign up for product announcements and analyst briefings as well as MSNBC, BBC, and local radio stations. Organizations use multicasting for distributing software updates.

15.8 The Downside of Push Technology

One of the major downsides to our current information age is information overload. With push technology, users are getting bombarded with information that someone (maybe someone other than themselves) thinks they need to do their job. Headlines, etc., designed to catch someone's attention, flash on the screen—interrupting thought processes. If disruptions are going to occur, the information contained had better be important and not junk mail (info-junk)—or even things that may be important to the user (ball scores, stock market quotes on their own stock, weather at their weekend vacation spot), but are not critical to the organization.

Corporations considering push technology must also consider network bandwidth. Organizations are already gobbling up every available bit of network bandwidth. Push technology is going to increase that requirement, which means decreased speed within the network.

Another point to consider is that BackWeb, Castanet, and PointCast actually initiate a request to the push server at a specified interval, a process referred to as client poll. Incisa, a truly push product, uses a multicast approach. Instead of requiring clients to check repeatedly for new content, Incisa sends one message that is routed to all recipients that are interested in the change. A client poll will have a much higher impact on bandwidth than will a multicast approach.

To avoid bottlenecks, administrators should establish push guidelines, limiting the use of external push sources or implementing internal caching to reduce the effects of pushing data to multiple employees. PointCast addresses these concerns with its I-Server. Targeted directly for intranet use, I-Server allows administrators to cache broadcasts on a local server and provides more control over resources.

In addition, channels are available free to users because of the advertising that comes with them. The advertising is in itself distracting, and it's eating up corporate bandwidth at the same time. Another reason these Web services are free is that the providers sell the demographic data, although they currently promise not to sell or rent their address lists to marketing companies.

One problem with the Web today is its relatively static content. Users want more action, more interesting content. Dynamic HTML can be used to update the content on a Web page with a live feed to multimedia and back-end databases. Java applets and ActiveX controls can be used to interact with server resources. More information on Dynamic HTML can be found in Sec. 8.2.3.

Acceptance of Microsoft's CDF standard will drive proprietary push products such as BackWeb's Channel Server, Marimaba Inc.'s Castanet, and PointCast Inc.'s PointCast Network to the Web. This will make it more difficult to stop push services from getting past the corporate firewall. Previously, filtering the proprietary traffic of specific push products such as BackWeb, Marimba, and PointCast was relatively simple. But with Microsoft Internet Explorer and Netscape Netcaster, the only way to stop pushed information will be to block all inbound HTTP and HTML traffic, since these two products will use this protocol and document standard to transport data through a firewall to an E-mail system.

Organizations can filter out specific addresses or implement a strict push policy, but these solutions limit users' ability to gather information that they really do need. Addressing these concerns are the companies that offer search engines, such as Verity Inc. and Open Text Corp. Future releases of search engines will make push technologies smarter by providing automated data delivery and advanced filtering based on very specific user profiles.

15.9   Is Push Necessary?

Push technology has found a new friend—the corporate intranet. Push on an intranet is a reliable and cost-effective method of communicating company information to employees. E-mail works fine, but messages are overlooked in the daily "pile" of unread mail or often deleted without even being read. Push products get around this by regularly broadcasting an announcement or a pointer to a stable information source. But push products will work only if they're used, and employees won't use them unless they are as easy to use as E-mail.

Organizations need to compare these message-oriented push servers to their existing push technology, E-mail. Groupware and/or E-mail is most likely already used to distribute time-sensitive information and files. Except for their ability to act as proxies for Internet news feeds, push servers that simply broadcast messages may be redundant for most networks.

Keep in mind that push is not a new idea. E-mail can be considered a push technology. Internet mailing lists are a rudimentary information push. However, push is here to stay, with its inclusion in the next generation of Web browsers. The challenge is how to make push technology work to benefit the users within the organization's network.

Push technology will certainly affect Web browsers and users' desktops. Early indications are that it will enhance the experience. Desktops are becoming more Web-centered. Getting information provided by the Web rather than surfed off the Web is a natural next step.

This first generation of products shows much promise. For many organizations, the standard E-mail system will continue to suffice for broadcasting information and files across the network. The infrastructure is in place, and users are used to its broadcast context already.

Push services oriented toward pushing information into specific applications, such as delivering statistics into an Excel spreadsheet, or, as with Marimba's Castanet, pushing the applications themselves, will begin to add value to an organization. And as push becomes integrated into the next generation of Internet clients from Microsoft and Netscape, organizations will have more options in customizing push capabilities.

Push systems should not be the focus. They are just a tool, not a product or a finished application. They should be used to provide the right information to the right user. They should not be focusing on the mere delivery of the latest headlines. Organizations should begin to think outside the box. Push technologies can be used to tie together such diverse sources as live audio and video feeds, Web-server content, and corporate databases. The possibilities are endless.

16

Delivering Push Technology

Push architecture is similar to Web architecture—there are clients, servers, and a network. However, for client/server and Web applications, the user initiates the actions. In push applications, the server initiates the action and sends the results to the client software.

16.1   Push Client Software

Push client software lets Internet and intranet users customize delivery of news and information directly to their desktops from a variety of sources. Push client programs don't really "push." Push suggests that clients are notified automatically whenever new content is available. Currently, only Incisa, from Wayfarer Communications, comes close to that model because it uses a proprietary protocol to identify new content. Most push clients use a subscription model: Clients poll the content servers at predefined intervals to see if new content is available.

Push clients run in background and automatically download information to the user's desktop. They are designed to deliver news and information but some can also deliver multimedia or software. Users customize the software to deliver only certain channels. When new information is delivered, push clients typically notify users through E-mail but could also alert a user by playing a sound, displaying an icon on the desktop, popping up the application, or displaying headlines on a screen saver, wallpaper, or a scrolling ticker tape.

While the graphical user interfaces may look similar, the products are somewhat different and definitely proprietary. Some products use HTTP (Hypertext Transfer Protocol) to download content index files; some also intercept HTTP requests between the browser and the TCP/IP stack. PointCast does not offer users a "linked" connection to the Web. Instead, hot links within PointCast material bring up a browser and take users in real time to the appropriate Web page. Castanet typically sends data in HTML (Hypertext Markup Language) format, which allows users to view channel content through a standard Web browser.

16.1.1   BackWeb's Client

BackWeb software (www.backweb.com) can deliver multimedia content for a richer presentation of information, including animation, as can Castanet from Marimba Inc. Although BackWeb's client, server, and developers' tools offer a complete solution for content providers and users, they use proprietary formats.

BackWeb's Infopak information-delivery programs must be downloaded every time a user gets new information. These programs are typically 100 to 400 kbytes in size, which can put a strain on disk space and network bandwidth. To lighten the load on LANs, BackWeb suggests using its "polite agent," which delivers packets only when the LAN or machine isn't busy and uses the more efficient, less demanding User Datagram Protocol (UDP) instead of HTTP to deliver content. To conserve disk space, BackWeb lets users set per-channel disk-space quotas.

The BackWeb client runs as an icon in the system tray, polling and gathering data from BackWeb channel servers. A click on the icon brings the client to the foreground, showing a list of new items, called InfoPaks, received from the servers. When an item is double-clicked, BackWeb launches that InfoPak. You can also set BackWeb to launch InfoPaks automatically when it receives them or to change your screen saver on the fly (if you receive an InfoPak that contains appropriate content). Any InfoPak can be designated as an alert, called a Flash, and the client software can be configured to display Flash InfoPaks as soon as they arrive.

An InfoPak can contain text, sound, graphics, URLs (uniform resource locators), HTML code, screen savers, bitmaps, or even executable files (although BackWeb can't automatically launch executable files). This ability to transfer any type of file makes it possible to use BackWeb as a simple software-distribution system. The software-distribution capability of BackWeb is very different, however, from that provided by Marimba's Castanet. Unlike Castanet, with its active Java applications, BackWeb just distributes files.

BackWeb's InfoPaks are essentially scripted packages of multimedia content, similar in spirit to Macromedia's Shockwave program. Using BackWeb's scripting language, Java applets, graphics files, audio files, and wallpapers can be packaged into an Infopak.

BackWeb charges content providers on a per-user basis and provides the client software free. Content providers purchase the BackWeb server and are charged a per-subscriber fee. The BackWeb server is used both by content providers and within corporate intranets, where specialized channels can be created to deliver information companywide.

16.1.2   Incisa's Client

Incisa from Wayfarer (www.wayfarer.com) offers a great deal of control for corporate users who want to deploy push on an intranet. Incisa is unique in that intranet users stay connected to the local Incisa server. Incisa uses a proprietary protocol that runs on top of TCP/IP to maintain a stateful connection between client and server, so that Incisa can push the latest information directly to users instead of polling sources periodically.

The Incisa server uniquely maintains a user and group database, which gives the administrator more leverage to customize user or group desktops. Administrators can create groups that should receive certain content and can then add users to the group.

Incisa's GUI client delivers headlines using reusable Shockwave animations combined with custom graphics and sound. By reusing the same Shockwave animations with different headlines, Incisa preserves bandwidth, since Shockwave animations don't have to be downloaded with each new headline.

Incisa's administrative interface lets administrators create installable clients with configuration options preset and locked down to desired settings. Administration of users' desktops is easier than with push clients that users must manually configure.

The Incisa client is a small, movable window that continuously displays incoming Incisa messages called HeadLinks. Unlike all the other products mentioned here, Incisa doesn't really have a concept of separate channels on the client. Each HeadLink appears for a few seconds in the main window, and then the next one takes its place. A user can scroll through a list of recent HeadLinks to find messages of interest.

HeadLinks are delivered as Shockwave animations, and Wayfarer provides several preconfigured Shockwave scripts for creating internal HeadLinks. Each HeadLink contains a heading, a few lines of text, and an optional URL.

16.1.3   Castanet's Client

Marimba's Castanet (www.marimba.com) is unique in that it wasn't designed to deliver the latest news; it was designed to deliver Java software—although Castanet can be used to download channels that are Java applets, applications, and even HTML pages. Marimba's Castanet suite of push products—the company's Tuner push client, Transmitter push content server, and Bongo authoring tool—can be used to deploy applications inside and outside corporate intranets.

Once a Java application is downloaded to a desktop, it is maintained and updated via the Tuner. Bongo makes it easy for Java developers to deploy Java applications or applets via a publishing wizard.

As a Java application, the Castanet Tuner lacks some of the bells and whistles of the other push clients because the Java application programming interfaces (APIs) don't yet support that functionality under Windows 95. For example, there are no screen-saver, wallpaper, or ticker-tape notifications of new content.

16.1.4   PointCast's Client

The PointCast client offers solid delivery of headlines and news. On the desktop, PointCast (www.pointcast.com) offers a pleasant user interface (if the user doesn't mind the advertisements), with graphics, text, and animations. For viewing articles, a user can use the built-in browser or have all content handed off to his or her own browser. PointCast's browser is no match for Internet Explorer or Navigator, since it doesn't support Java, scripting languages, or advanced HTML. It is designed just to display basic news pages, using HTML and graphics.

PointCast Network broadcasts news stories—business updates, stock quotes, news headlines, etc.—to any user who has PointCast's client software, which is downloadable for free from PointCast's Internet site. If the user has a continuous connection to the Net, the items are automatically broadcast to the computer. If the user has dial-up access to the Net, the updates are downloaded when the user goes on-line. PointCast channels are category-based—finance, sports, etc. Any changes within the Web pages of the selected category are sent to the user's desktop. If a user chooses finance, any changes to finance-related Web pages will be delivered—even if the only area of finance the user is really interested in is the prime rate.

The PointCast screen saver flashes the latest items on the monitor when the computer is idle. If a user wants more information about an item that appears on the screen saver, a click on the headline brings up the article, which has been downloaded to the PC. To conserve hard drive space, a user can tell the software how much hard drive space can be had for PointCast stories. The software overwrites old stories when it needs space.

PointCast does not offer users a "linked" connection to the Web. Instead, PointCast material include hot links that bring up a browser and take users to the appropriate Web page in real time.

PointCast Inc. is a channel aggregator, which means that PointCast can easily offer content personalization and careful control over what's being offered. However, it also means that users are limited to the channels that PointCast offers, currently 29.

16.1.5   Transceive

Caravelle Inc.'s Transceive (www.caravelle.com) uses a simple HTTP-based approach for its implementation of push. Its Receiver, Producer, and Publisher products can be deployed quickly with existing Web clients and servers.

Caravelle's channels are simply Web pages, and the vendor uses HTTP to determine if a channel has been updated by checking the page's last modification date. Creating channels is trivial because a channel is just a Web page. By the same token, the Caravelle channels won't give the user specific information packaged in specific headlines; the user just gets the target Web page.

Caravelle's Producer, like Incisa, can be used to create customized Receivers with the options locked down, making it easy for an adminis- trator to set up new users and control what is being offered. The Caravelle Publisher monitors files on the content provider's system for updates and then posts the modified files to the organization's Web server so that Receiver clients can access the new information.

16.2   Push Technology on the Desktop

Push technologies are being used by Microsoft Internet Explorer 4.0 and Netscape Netcaster to change the way users receive and use information from the Internet.

Once push is integrated with Web browsers, push delivery will become a common part of a user's information-gathering experience and will complement the browser. Internet Explorer includes the ability to connect to PointCast channels as well as other channels. Netcaster includes Marimba's Castanet Tuner. But the client portions of products such as BackWeb, Castanet, and PointCast continue to provide features beyond the basic push technology included with the browsers.

The new push desktops combine the features of existing push prod- ucts into a single client. However, these two major players are approach- ing push technology differently. Microsoft brings the Web to its popular Windows desktop by embedding pieces of the Web browser throughout the operating system and user interface (the basis of Microsoft's current court battle with the Justice Department). Netscape Netcaster brings the desktop into its Web interface Navigator. The Netcaster desktop allows a user to access Web sites and E-mail systems and perform other Internet tasks. It also provides an interface to applications and files on the local hard drive.

Microsoft's push interface is grafted directly onto the Windows desk- top. Netscape has positioned its offerings as crossware, cross-platform desktops. Netscape's Constellation looks more server-based than Active

Desktop, so users can log onto their desktop machine from any computer.

Notification becomes an important feature of these desktop software offerings. A smart Internet client should not only notify a user when pages of the user's favorite Web site have changed, but should download the changed content automatically even when the user is disconnected from the network.

Just like the strictly push client software, the Microsoft and Netscape desktops deliver access to channels.

16.2.1 Push in Microsoft Internet Explorer

The most powerful push technology in Microsoft Internet Explorer 4.0 is its adoption of channels. Webmasters can build Microsoft-compatible channels using ActiveX, Dynamic HTML, HTML, and Java.

The other push technology in Internet Explorer 4.0 is the notion of subscriptions to Web sites. With subscriptions (which are free), Web sites send Internet Explorer users content automatically. Internet Explorer can download subscribed Web-site data unattended, and the browser notifies the user when new content becomes available via an icon on the Favorites menu next to pages that have changed. Users can also choose to have Internet Explorer notify them of Web-site changes via E-mail using Outlook Express, Internet Explorer's new integrated mail and news client.

Microsoft's Active Desktop will integrate a broadcast-oriented interface into the Internet Explorer 4.0 browser and the next version of Windows—Windows 98.

16.2.2 Push in Netscape Netcaster

The Netscape browser-and-desktop Netcaster is primarily a communications client. It is included as a component in Netscape Communicator, which is discussed in Sec. 9.2.3.

The Channel Finder in Netcaster allows users to select from one of two sets of preestablished default channels rather than allowing users to decide what Web location they would like to visit. In addition, any existing Web site can be viewed as a channel. Favorite channels can be anchored to a user's desktop, creating a full-screen information-centric workspace called a "webtop." (A webtop is the Web equivalent of a Windows desktop.)

Netcaster can notify users about changed content in favorite Web sites. When a site changes, the user sees a message on the desktop (or Homeport, in Netscape's terminology). A notification engine is built into Netcaster to handle these real-time messages and to display them to the user.

Web broadcasting in Netscape works with Dynamic HTML, HTML, Java, and JavaScript through standard HTTP servers. There are no functional changes required on the server side or on the client side, or in the Web software, the firewall, or other systems.

Netcaster handles push content in a number of ways. LiveSites are the Netscape equivalent of channels. A Netscape LiveSite is much like an existing Web site, but its content is marked for broadcast. Netscape also plans to include the Castanet Tuner in Netcaster.

Netcaster also has something called InfoStream, which is areas of the desktop that can deliver small blocks of information in real time. These pagelets can also be used to display pushed content.

Netscape's Directory Server 3.0 is an important part of Netscape's push technology. Directory Server stores user profiles and permissions and assists in the management of who is allowed to push content where. Administrators can provide authority to subscribe and directories of who can subscribe to what within the company. Such directory services allow organizations to manage customer interactions as well. Directory Server 3.0 was released in late 1997. Later versions of Directory Server in the Apollo and Mercury releases slated for early 1998 will include a CORBA object-finder capability.

Netscape's Constellation is a Java-based interface that organizes the desktop into channels of push content. After a user selects the sources of data that should be pushed to the desktop, such as Web-based news or financial sites, Constellation will regularly update the data from the Web directly to the desktop. Netscape's Mercury extends this push metaphor with an agent-based technology, code-named Compass, to filter on-line information automatically. A server suite, code-named Apollo, will feature push content and replace the currently offered SuiteSpot. Constellation, Mercury, and Apollo are due in early 1998.

16.2.3   Sun's HotJava Views

Sun Microsystems' HotJava Views is a Java-based GUI (Sun calls it a "webtop") for its JavaStation network computers that includes push-enabled features along with its core E-mail, calendaring, and Web access tools. JavaStation administrators define a profile of what software and content users would be interested in and have it pushed to the users' desktops. The JavaStation architecture stores users' computing desktop parameters on a server so that individual desktops are available from any machine.

16.3 Push Server Software

There are major differences among the four major push server offerings: BackWeb, Marimba's Castanet, Wayfarer's Incisa, and PointCast I-Server. Marimba's Castanet, especially designed to deliver and maintain applications, offers a unique management system. When it updates an application, Castanet sends only the data that has changed, which minimizes network traffic. Castanet is built around Java and is a good choice for networks using network-based Java applications. A future version of Castanet will support non-Java applications. BackWeb's software distribution is a simple file download.

The three other major players focus on distributing messages and Web pages. PointCast I-Server basically provides a proxy server for the service's news and information content and allows an organization to set up its own PointCast channel. BackWeb has more diverse content-delivery capabilities, but at the price of more complexity. Incisa is designed for intranets and includes no advertising as PointCast does.

Some delivery services that use push technology, such as Digital Bindery, allow the user to specify only preferred Web sites. A user enters the URLs of the Web pages of interest. Digital Bindery monitors those sites and transmits changes once a day.

As illustrated in Fig. 16-1, the ideal push architecture should be augmented by filtering, indexing, and directory services to facilitate the management of the content and the push channels.

Figure 16-1
Additions to Standard
Push Implementations

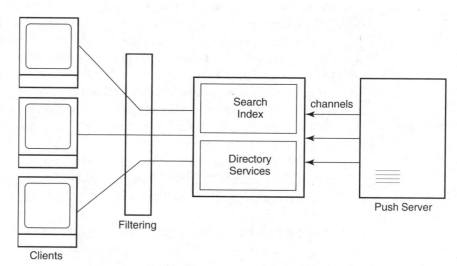

- *Filtering* allows the user to select content subsets of a single channel or a subset of the entire number of channels available to that user.

- *Indexing* of channel content allows the user to search pushed content for related content that may have appeared earlier.

- *Directory services* let IT managers set user profiles to control which channels the user can subscribe to, which secure channels the user is authorized to view, and which desktops have the power to assign channels to other desktops.

16.3.1 PointCast I-Server

PointCast, the founder of push, offers I-Server and its administration tools that provide HTML-based information broadcasting capabilities for the network administrator. However, currently I-Server 1.2 can deliver only HTML pages and links, compared with Castanet's Java applications and BackWeb's multiple file formats. In addition, I-Server doesn't allow network administrators to shut off channels or their associated advertising. PointCast is considering future versions of I-Server that would allow blocking of channels.

In addition, there's no way to limit or direct content flow to groups of intranet clients because the PointCast I-Server has no knowledge of which users are connected.

The I-Server, which runs on Windows NT Server, acts as the conduit between the PointCast clients on an internal network and the Internet-based PointCast Network itself. Information channels feed to the server, which then sends the content to the requesting clients.

To receive PointCast's news and information feeds, the I-Server requires an Internet connection. When used without the I-Server, PointCast can occupy considerable bandwidth on companies' Internet connections. The I-Server's caching capability entirely eliminates requests for the same information by multiple clients, reducing Internet-bound traffic. The server will not reduce internal LAN traffic, but most sites will find that traffic generated by PointCast clients accessing an I-Server has a minor effect on network capacity.

Besides the variety of commercial news services provided through PointCast—which include CNN, *The New York Times*, and Reuters—the I-Server includes an important addition: the Corporate Channel, which lets an organization send its own HTML content over PointCast. This channel can be named anything the organization wishes, and it can be associated with a corporate logo. The channel then appears on PointCast clients as the last channel on the selector.

Administrators can create any number of subchannels of the Corporate Channel to organize content. Typical subchannels may include human resources, sales, or any subject that applies to the organization. Each subchannel contains the article title and URL for the HTML file. The newsticker editing interface includes easy add and remove capabilities to indicate which items to include in the ticker, as well as a feature for broadcasting information flashes and headlines in near real time.

PointCast recently announced PointCast Connections, a component of its next version that will allow anyone to publish a PointCast subchannel using Microsoft's proposed Channel Definition Format (CDF). Also, the final release of Microsoft Internet Explorer 4.0 includes the ability to receive PointCast content.

Currently, PointCast has no developer kits or tools available, such as those provided by BackWeb and Marimba, though PointCast plans to remedy this in a future release. The PointCast client viewing template and interface are unalterable. Unlike Wayfarer's Incisa, I-Server doesn't provide tools to access back-end databases. But Domino.Broadcast for PointCast (developed jointly with Lotus) allows delivery of HTML-based Domino content via the PointCast Corporate Channel.

The I-Server's administration tools are accessible via a Web browser or Windows console, and it lets an administrator remotely track multiple I-Servers. I-Server has extensive built-in transaction and error logging, along with accompanying viewer-analysis tools for summaries or full reports.

16.3.2 BackWeb

As its name implies, BackWeb works like a Web server in reverse. The BackWeb client allows users to choose which channels they want to receive. These can be a mix of public channels available on the Internet and private channels transmitted over the corporate intranet.

BackWeb can transfer information in the background, using a "polite agent" that stops while the user is actively browsing or downloading material from other sites, then resumes when bandwidth is free again.

The BackWeb Channel Server allows an organization to set up and manage its own channels. The BackWeb server consists of three parts: the server engine itself, the server console, and an editor for BALI (the BackWeb Authoring Language Interface). The server console program, which runs on any Windows 95 or Windows NT PC attached to the network, allows administrators to add and delete InfoPaks and provides tools that are used to control the behavior of InfoPaks. For example, administrators can control the time, date, or number of times an InfoPak is displayed, as well as tailor InfoPaks for users who have specific computer types, display resolutions, or multimedia capabilities.

16.3.3   Incisa

Wayfarer Communications Inc.'s Incisa is a corporate data broadcasting tool that allows managers to send carefully filtered news, messages, and Internet URLs to Incisa users' desktops. Compared to the other push servers, Incisa's lack of a channel metaphor limits its usefulness. PointCast is an advertising-supported service aimed at the general public, whereas Incisa targets the corporate workplace. Unlike PointCast, Incisa's news and stock feeds don't contain advertising, and Incisa provides sophisticated data filtering tools for control over what each user sees.

Like that of BackWeb, the Incisa server has a minimal management interface. Management chores are performed with the Incisa Reporter program, which runs on any network PC. Using Reporter, an administrator tells the server what data it is to collect and which Incisa groups are to receive it. Reporter contains a very thorough suite of data-filtering tools which allow an administrator to route specific types of information to specific groups of Incisa users. For example, management users might be allowed to see the Reuters business news feed while other users are restricted to internally generated information.

The Incisa server automatically creates HeadLinks for each news item received from the news feed. Internal HeadLinks are created using the Reporter program. The administrator selects which animation to use; types in a heading, a few lines of text, and an optional URL; and then selects the Incisa group or groups to receive the message. The new Head-Link is placed in the outgoing message queue along with additional Head-Links from the news feed. If the Urgent option is checked, Incisa places that HeadLink at the top of the list to be seen by the Incisa client.

Incisa can import live data from several sources, including Microsoft Access, Microsoft SQL Server, and Oracle databases. The Incisa DataBridge provides an Open Database Connectivity (ODBC) connection to these databases and can automatically create HeadLinks on the fly based on new information.

Unique among these products, Incisa uses a persistent connection between the server and clients. Most push server software uses a client-polling technique in which the client software periodically checks for updates on the server. Incisa's QuickServer technology allows for faster delivery of HeadLinks, since the client doesn't have to reestablish a connection with the server each time it receives a HeadLink. Also, the server needs to send a HeadLink just once over the network, instead of delivering it every time an individual client requests it.

16.3.4   Castanet

Though message and Web-page delivery is the focus of the other products, that's just one application of Marimba Inc.'s Castanet. Castanet's focus is on distributing Java applications. With Castanet Tuner, a user can tune in to a channel that broadcasts news updates or to a Castanet channel that sends out a product catalog or a word processor written in Java. At present, Castanet channels are able to distribute Java applications and HTML pages.

Castanet's major difference is its intelligent methods of updating the channels it broadcasts. All files are sent using a single TCP connection, and only modified parts of files are sent across the network. Channels that share Java code can also share updates, so that if several channels need a new version of a certain Java class, the Tuner downloads it only once.

Castanet channels can also distribute HTML pages. Administrators can create a channel that contains an entire Web site so that users can keep a copy locally and browse off-line.

Castanet Transmitters, the server portion of Castanet, deliver channels to Tuners over the Internet or the company's intranet. In addition to the Transmitter software, servers can also run Castanet Proxies, which will periodically check the original servers to make sure channels are up-to-date, and Castanet Repeaters, which will mirror channels across multiple Transmitters. But unlike mirrored Web or ftp sites, a Repeater automatically checks to ensure that these are the latest updates. Repeaters can also perform load balancing to redirect client requests to a Transmitter that is less busy or geographically closer to the client.

To set up the software, an administrator specifies an installation directory and some simple configuration data. A Transmitter and a Tuner can be set up on the same computer to test channels before they are published. When a user subscribes to a channel, the Castanet Tuner downloads the application and launches it all in one step. Users can update the channel in the Tuner manually. Alternatively, the author of the channel can specify when to check for updates.

16.3.5   WebCast Studio

Astound Inc.'s WebCast Studio (www.astound.com) allows an organization to create channels that pull content from existing Web pages on the fly and does not require a server-side component. WebCast is a pull service that sends updates when users launch the player. In contrast to Point-Cast I-Server, which focuses on corporate intranets and pulls its content from corporate servers and the PointCast Data Center, WebCast Studio's server-free approach allows Web developers to use content from their own Web sites, other Web sites, or even HTML pages that are not live on the Web.

WebCast Studio can be used to create a single flash screen or slide show viewable with the free player. The WebCast Studio authoring module supports a variety of special effects. It requires no HTML coding, and access to all the functions is through a series of menus, buttons, and dialog boxes.

WebCast Studio divides into three views: channel, content, and template. The channel view is used to set up the initial channel (logos, text boxes, buttons, etc.), and the content view is used to identify the content that goes on the pages. The template view allows a user to create a template for the entire channel.

WebCast Studio comes in three versions: Personal, Professional, and Commercial. Personal and Professional are targeted at small and medium-size Web-site designers, and there is no limit on the number of channels that can be created or on their complexity. The commercial version is for large Web sites with lots of ads.

16.3.6   Transceive

Caravelle, Inc.'s Transceive acts as a proxy server, monitoring selected Web sites, collecting information as Web pages are updated, and distributing the new information to users across a corporate intranet. The product can also broadcast documents across an intranet or the Internet.

Transceive copies pages from selected Web sites to servers inside the firewall, so that users can access the information contained on the site without going out on the public Internet. Transceive does highlight changes to documents it collects when it delivers the updated document. Its push technology is similar to that used by BackWeb and Wayfarer.

The product consists of three components:

■ The *Receiver* is a client-side tool used by users to create their own channels that contain the sites and pages they want. Because there is no need for specific server software, users can check any site the viewer can access. The Receiver can then go out and pull in updates at whatever intervals a user has set. Users can see the updates they want and archive changes over time.

■ The *Producer* can be used to lock predefined Web addresses into the Receivers. Receivers can then be distributed directly to a specific audience, thereby establishing a private network as either an intranet or an extranet. Just by posting new information to a Web server once, users can instantly connect to a plugged-in or targeted audience. The Producer can save valuable network resources by taking over browsing and searching duties, as well as publishing in one step.

■ The *Publisher* essentially combines all the features of Receiver plus the capability to publish updates in HTML format on the server. This capability enables workgroups in an organization with common interests to tap into channels without having to duplicate the effort.

Because Transceive accesses specific sites, it is ideal for users with limited, easily defined sources of information, but falls short for users who need unrestricted access to the Internet.

16.4 Push Technology in Groupware

At its most basic level, groupware supports broadcast information via E-mail. Users then act on received information directly in the groupware client. A groupware application could include intelligent information-routing and approval rules. Most standalone push software products, in contrast, are designed as one-way broadcast media.

Like push software, groupware can analyze information and distribute it to individual users. Groupware products such as Lotus Domino, Microsoft Exchange, and Novell's GroupWise already include many push features for distributing information throughout an organization. Domino can be used to create powerful agents that can periodically check a collection of documents and perform actions on what they find. Exchange and GroupWise use rules to perform similar kinds of tasks. Agents or rules can be used to glean information from a variety of sources based on a set of criteria, such as a company name or news source.

Groupware products are used to collect information in databases and data stores. Exchange and Lotus Domino provide APIs to allow developers to manipulate data in the data store. This access allows a variety of data-driven applications. For example, news-wire stories can be converted into individual documents in a Domino database. Lotus Notes users can access these feeds directly from the database.

Most groupware products have evolved beyond simply forwarding documents from a database. Exchange, for example, lets users send E-mail with just a link to the actual data. Domino adds its newsletter feature, which periodically generates a single message with multiple links instead of dozens of individual messages. Groupware packages can allow an organization to create a push system without having to maintain separate push clients and servers.

Lotus has announced Domino.Broadcast, which will be able to publish Domino database information using a variety of push servers, including those from BackWeb, Marimba, PointCast, and Wayfarer.

Netscape offers its own groupware "suite" under the umbrella name Communicator. The client suite is composed of Navigator 4.0, Netcaster content delivery system, Messenger for open E-mail, Composer for HTML editing/publishing, Collabra for group collaboration, and Conference for real-time audioconferencing. A professional edition is also available which includes Calendar for enterprisewide calendaring/scheduling, AutoAdmin for central IS administration, and IBM Host On-Demand for 3270 terminal emulation.

16.5   Search-Enabled Push Software

Search-enabled push technologies, such as Verity's Search'97 IntelliServ, monitor a variety of sources of information—Web sites, news services, or any accessible document database—for changes that are relevant to a user's profile. The system doesn't require any proprietary client software. Information is received through a standard Web browser or can be delivered through E-mail as HTML. IntelliServ uses Microsoft's Active Server Page (ASP) technology to push updated information to a search page as it locates it.

Administrators assign rights to groups and users for various data sources and determine how many levels deep to search a Web site—two levels would result in an index of the main page and all pages linked from it.

User profiles are very specific with search-enabled push technologies. Rather than specifying a broad category, such as automobiles, users can be more specific, such as 4WD vehicles.

In addition, a weekly newsletter is "published" by changing an existing file that is monitored by a search engine. When the file changes, it is delivered automatically to users who have searched for just such information.

16.6 Push for Software Distribution

Push technology can also be used to distribute applications to multiple users. Castanet is designed to distribute and manage applications automatically. BackWeb provides software distribution but does not currently offer an entire framework for distributing and running programs, as Castanet does.

For many IT organizations, this aspect of push technology offers the greatest benefit. Toys "R" Us uses a custom IP multicast solution to distribute software updates; it takes four minutes to update 250 clients nationwide. Boeing is using a multicast solution to update productivity software on hundreds of thousands of desktop machines overnight.

16.7 Sample Applications

IT managers are adopting push technology to automate and streamline jobs that can be time-consuming and costly, such as distributing and maintaining software, triggering data processes, and publishing sales and inventory information to business partners.

Merrill Lynch   Merrill Lynch Inc. is using push technology as part of its new Trusted Global Advisor (TGA) system, to be deployed in 1998 to more than 25,000 users in 700 locations. It will provide financial consultants with news reports, portfolio management tools, investment tools, stock market information, client information, and historical data. Desktop Data Inc.'s NewsEdge on-line news service is used to provide consultants with customized news feeds about clients and companies they track. Historical analysis is pushed out of Merrill's database on a scheduled basis.

Fidelity Investments   Fidelity Investments uses push technology as an alternative to E-mail for notifying key employees of internal information that is currently scattered among numerous corporate databases. Using BackWeb, Fidelity is pushing financial, operational, and investment data from an Oracle Corp. database to analysts on a weekly basis. The system can also embed live links to intranet pages if users need additional information.

Ascend Communications   Ascend Communications Inc. uses Diffusion Inc.'s IntraExpress push system to reduce its reliance on paper-based information systems. Ascend, a remote access network vendor, is using IntraExpress to communicate with its 100 value-added resellers, and ultimately with its 1300-member field sales force. Ascend is also converting its newsletter into a weekly pushed edition rather than a paper newsletter every six weeks.

St. Luke Episcopal Hospital   In an effort to reduce the amount of paper that crosses the desks of employees, St. Luke Episcopal Hospital, part of the Houston Medical Center, deployed an intranet and an information push system based on Wayfarer Inc.'s Incisa server technology. Supporting more than 1500 users, the system sends out updates and news flashes about goings-on at the hospital. It's currently the front end for the hospital's intranet, but plans are being made to break up the information being sent, in order to target different groups' interests.

Fruit of the Loom   Fruit of the Loom runs its own internal, company-wide channel on the PointCast Network. Using PointCast's I-Server software running on a Windows NT server, all the PointCast Network updates from CNN and other content providers are downloaded and then broadcast to the PCs on Fruit of the Loom's internal network. In addition, Fruit of the Loom also broadcasts its own news and reports from within its firewall-protected intranet. Any internal information that is to be shared is sent to the Web team for broadcast. The push on the intranet has allowed Fruit of the Loom to share inventory level information with the sales force, so that they know what is in stock. Fruit of the Loom has installed PointCast client software on the PCs of its major wholesalers so that they can transmit information throughout the day.

AT&T Wireless Services   AT&T Wireless Services uses Netscape Communicator to send members of its Developer Program the latest technical data, sample code, and other development information. Developers download the subscriber software from the Web site and specify which information they want to receive. Then, whenever AT&T Wireless Services has a relevant product or news update, the Communicator publisher sends the information to the developers' desktops. Developers get the information they need and are not bombarded with information they don't need.

PART 5

Electronic Commerce

Electronic commerce (E-commerce) is quickly becoming one of the mainstays of business. The value of business E-commerce transactions will hit $120 billion by 2000, according to some optimistic analysts' assessments.

E-commerce is business conducted among trading partners over the Internet, intranets, or private networks. The term is used to cover electronic shopping, electronic data interchange (EDI), and transaction processing. It enables companies to efficiently provide products and services to a large and varied audience.

EDI was a first step toward electronic communication between trading partners. EDI has been used for over a decade in one form or another. With EDI, electronic standardized formats of documents are transferred over a private network between systems run by different companies. That concept has evolved into customers' using electronic storefronts to purchase products and services on-line.

E-commerce is usually consumer-to-business or business-to-business. Consumer-to-business applications—electronic shopping—use the Internet, and payment is usually by credit card or electronic cash. Electronic shopping requires the creation of an electronic storefront, usually through the use of a commerce server, also known as a merchant server. Users have "carts" to drop their goods into (or take them out of), and "checkouts" to settle the bill. Any Internet user can browse and order goods or services from the storefront's on-line catalog.

Business-to-business applications could be called EDI applications. These Web-based EDI applications are reengineering the fulfillment and procurement processes in organizations. Their users are restricted to internal customers and business partners, so users' identities are verified. Payment is usually predefined between the trading partners. Instead of requiring an expensive, private, value-added network (VAN), transactions fly over the Net.

In places such as Japan and Singapore, 15 to 20 percent of Internet users buy goods and services via the Web. In the United States and Europe, 25 percent of home Internet users and 20 percent of business users turn to the Web for purchases. In more than half the transactions, actual payment is made via the Web; three-quarters of those are business-to-business transactions. IDC forecasts that $223 billion worth of goods and services will be purchased over the Internet in the year 2000, with most of it business-to-business.

Uses of E-commerce vary. A regional grocery store chain no longer carries its own inventory. It rents shelf space to its suppliers and manages the flow of products through an electronic just-in-time delivery system linked to its checkout scanners. A bank has increased its fee revenues by providing electronic corporate payment services via the Internet to local merchants and major corporate clients, such as utilities and government agencies, thereby slashing their accounts payable and accounts receivable costs since there are no paper checks to be handled and processed. A major PC manufacturer has streamlined its supply chain by putting all its trading partners on-line. Paperwork is eliminated; inventory handling charges are slashed; and the ordering, manufacturing, and shipping processes are automated.

The constantly growing use of the Internet by consumers makes the prospect of widespread electronic commerce using electronic storefronts increasingly likely. Companies should begin to plan support for Internet transactions on their Web sites. Often-voiced fears concerning the Internet's lack of security, insufficient credit-card encryption technology, and inadequately designed database management system utilities are being addressed slowly. Banks are engaging in multimillion-dollar transactions over the Internet, credit-card companies have dramatically enhanced their encryption technologies, and many new software vendors offer Internet-based transaction-processing monitor capabilities.

So instead of leafing through a catalog and calling an 800 number, we'll sit in front of our screens, make the items we are interested in "move" and change colors, drop them in our virtual shopping cart, look around some more, click on Pay, and have the goods arrive in a few days. Our order will generate some activity in inventory levels, which will prompt an order to a supplier, which will generate an acknowledgement, a shipment, and an invoice. All this will be done electronically and automatically (where possible)—efficiently and effectively.

Welcome to the electronic age.

17

Electronic Storefronts

We are used to multitiered distribution channels. Manufacturers pass products to distributors, who provide them to resellers, who configure the products and deliver them to end users. A few manufacturers sell and deliver directly to customers. Many vendors sell directly to their largest accounts, but configuration, delivery and installation are often handled by third parties. The biggest cost in this multitier distribution process is keeping the entire pipeline filled with inventory. And the biggest challenge is to have just the right amount of inventory in the pipeline.

Enter the Web. Consumers can go on-line, find products that meet their particular needs, do comparison shopping, and find articles that discuss the tradeoffs among features, price, and availability. The consumer can order directly from many Web sites, with the order filled within a specified period of time—often by the next day.

Figure 17-1
The Electronic Marketplace

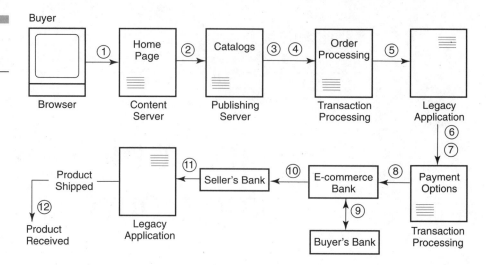

How does an electronic marketplace function? It has the same components and basic processes as a "real" marketplace does. The only difference is, as illustrated in Fig. 17-1, that it takes advantage of technology and is open 24 hours a day, 7 days a week.

The following items correspond to the circled numbers in the figure:

1. The buyer logs on to the marketplace—the Web page that provides the stepping-off point for the buyer.

2. The buyer searches for products. Some implementations also list advertising, news, and specials.

3. The buyer selects a product(s) from an on-line catalog. The content of this catalog can be generated dynamically based on the user's profile. The buyer puts the product(s) in an electronic shopping cart while browsing through the catalog.

4. The buyer orders the product(s). The storefront software pulls prices off data sources, calculates taxes, and totals the order.

5. The storefront software sends this data to the seller's legacy application for the actual processing of the order. In effect, the software sends in the order for the items the buyer has picked.

6. The seller confirms the order to the buyer.

7. The buyer pays for the products ordered.

8. The seller then takes the purchasing information that the buyer supplied and sends that payment information to the E-commerce bank.

9. The bank verifies the creditworthiness of the buyer (or validates the credit-card number).

10. The E-commerce bank lets the seller's bank know that the transaction has been approved.

11. The seller's legacy systems process the order (fulfillment) and ship the product(s).

12. The product is received by the buyer.

An electronic storefront (also called an electronic store) is a Web page that consumers access to perform the same tasks they might perform in a physical storefront. The consumer should be able to (ideally) look at the merchandise, price out the goods being offered, get information about the items being reviewed or even items previously purchased, and ultimately be able to actually purchase an item. In some cases, this interaction is in the form of a negotiation—buyers find the product they want and submit an "offer to buy" to every seller of the item; the sellers receive this offer via E-mail and log their responses; and the buyers compare prices and place on-line orders.

In order for the Internet to be the connection between the consumer and the merchant to accomplish all these tasks, organizations have to provide complete and secure order fulfillment (the "back-end") over the Internet.

17.1   Electronic Storefronts

Just as a merchant has many choices when opening a store in the real world, so do merchants who use the Web. Just as a traditional store benefits from careful attention to details when the store itself is being laid out, so does an electronic store. Some of the details that must be determined are illustrated in Fig. 17-2 and explained below. Just as traditional merchants can decide on low-cost fixtures and equipment, an electronic merchant has choices ranging from an all-in-one, entry-level Internet-storefront package to a high-end commerce product. Just as traditional merchants decide what payment options they will offer, an electronic merchant must decide how it will accept payment: digital cash only, or digital cash and credit, and if credit, how the merchant will secure credit information.

Browser   Shoppers can access the electronic store using any browser (although the storefront interface may look better with one browser than with another). If the shopper expects to make purchases, for security, he or she should use a browser that supports SSL (Secure Socket Layer), S-HTTP (Secure HTTP), or SET (Secure Electronic Transactions).

Figure 17-2
Architecture for an
Electronic Shop

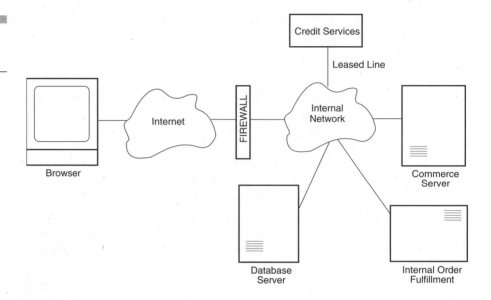

Firewall The firewall should be in front of the commerce server for security reasons, "protecting" it from unauthorized access. Many electronic storefront implementations also put a firewall in front of the database server for an additional layer of security.

Credit Services The commerce server needs to have access to credit verification services. These services might include both credit cards and digital cash payments. If the commerce server software supports SSL, S-HTTP, or SET, the connection can be through the Internet; otherwise for security it should be a direct connection to verification services.

Commerce Server The storefront software, the transaction service software, and possibly the data itself all sit on the commerce server. For better performance, many electronic storefronts keep the data on a separate server.

Database Server The data required to accept and place an order, as well as fulfillment data such as product and inventory data, is maintained on the database server.

Internal Order Fulfillment These legacy systems are often referred to as the back-end systems. The order has been accepted; availability of inventory for the order has been checked; the status of all items in the order has been determined (ship or back-order, for example); an acknowledgment has been sent; and internal applications such as inventory have been executed. Fulfillment is the last step—the process of actually filling and shipping the order.

17.1.1 Storefront Components

A Web storefront has the following components (most have counterparts in traditional storefronts):

- A *shopping cart system* that tracks user purchases as the user navigates a site and checks out purchases
- A *payment processor* that processes transactions and payments
- A *database server* that handles transactions and maintains data on store items, customers, and so forth
- A *catalog manager* that uses software to add items to the database, matches icons and graphics with store items, and presents catalogs for viewing
- An *accounting manager,* which is software that takes information from the shopping cart and passes it to corporate accounting

A merchant bank account instantly authorizes credit-card transactions and accepts deposits of credit-card receipts from a business or consumer. For a retail business to accept credit-card payments, it must have this type of account.

The Web-browsing shopper sees the storefront simply as a Web site whose purpose is to sell products. The store is usually structured hierarchically, with the home page acting as the point of entry. Just as a real store is divided into departments, electronic stores can be designed with product category screens to make it easy to locate products. Instead of asking someone to find a given item, the consumer queries the database for a given item using a screen.

17.2 Cybermalls

A cybermall is a shopping mall on the Internet. A cybermall links a home page to hundreds or thousands of on-line storefronts. The cybermall generally handles the financial transactions for all the merchants, so that a customer does not have to enter duplicate name and address information at each store. Items can be placed into an on-line shopping cart and paid for all at once by credit card, ecash, or another digital money method.

The iMALL (www.imall.com), one of the biggest Web shopping sites, has developed a transaction system that enables merchants to sell goods over the Web without worrying about the mechanics of electronic retailing. Anyone with as few as ten products to sell can pay a start-up fee as low as $100 to become commerce-enabled. All the organization has to do is add a "Buy Now" button to its Web page and a link that goes back to the iMall, where all the database and credit-card authorization takes place. With the iMall's new "Order Easy" system, shoppers can take a virtual shopping cart all around the Web, adding any desired items to the cart if they see a "Buy Now" button on a site.

iMALL includes dozens of commerce categories, search capabilities, advertising banners, and Cybercash secure transactions. The iMALL was designed to revolutionize the way people acquire knowledge and information, conduct business, and interact with one another. Recognized as a top commerce site by industry experts, trade publications, and leading Web directories, iMALL defined the standard for electronic shopping.

Attracting shoppers on the Web can be a risky business. And once there, Web shoppers are extremely finicky about what they will or won't buy—even for brand names, as IBM found out. IBM shut down its World Avenue cybermall after a year of unsuccessful attempts to build user interest in on-line shopping. The company shifted its focus from aggregate electronic commerce initiatives such as cybermalls to providing development and hosting services for on-line retailers.

MCI's electronic shopping mall, introduced in April 1995, closed its virtual doors a year later. MCI reportedly spent $10 million on developing the mall but closed it because of poor revenue. MCI feels that the mall failed in part because catalog vendors put only partial inventories on-line and insisted on flashy graphics which slowed the download time of the Web pages.

While larger companies with prominent brands have shifted away from cybermall venues to establish their own individual sites, smaller companies can benefit from piggybacking on a brand like AOL and Prodigy to draw visitors to their sites. A classified ad with a commercial on-line carrier or electronic mall generally runs less than $20 a month. If the ad permits linking to the company's Web site, add another $20 a month.

17.3 Payment Security

One of the attractive features of the Web has always been its openness. As wonderful as that is, it complicates E-commerce because the Web is not secure. Security must be handled by individual sites on the Web: Secure the server at all (well, reasonable) costs.

Three fundamentals of Web-server security are fortification, authentication, and encryption. Firewalls provide the fortification. Security protocols can handle the authentication of strangers and encryption of messages. A secure Web commerce server should support one or more of the major security protocols. The security protocols emerging on the Web are Netscape's Secure Socket Layer, NCSA's Secure Hypertext Transfer Protocol, and Microsoft's Private Communications Technology (PCT), as well as the new Secure Electronic Transactions protocol. Web browsers and servers are expected to support all the popular security protocols as they become generally accepted.

Security protocols are discussed in more detail in Sec. 11.4.

17.3.1 Encryption Indicators

Microsoft Internet Explorer and Netscape Navigator both let a user know when the Secure Socket Layer is being used to pass encrypted information between the browser and a Web site. This is especially critical when any part of an on-line transaction involves digital cash or credit-card information. Navigator shows a key icon, and Internet Explorer shows a yellow padlock. Both provide a pop-up dialog box when switching between secure and unsecure modes. Users can turn this warning off if desired.

17.4 Electronic Wallets

Electronic wallets are virtual wallets that hold credit-card numbers and expiration dates, as well as shipping information and/or digital money. Electronic wallet software resides on a user's own hard drive—and we may see it on smart cards before too long. A user needs to link the wallet to one or more existing bank accounts or credit cards to be able to "deposit" money into the wallet. The wallet software also provides the interface for managing coins and handles the security required to transmit financial data over the Net. The wallet is password-protected to prevent others from spending the money.

When a user wants to buy goods or comes across an entertainment provider or pay-as-you-go information site, the user clicks on a Pay button. The wallet opens, the user clicks that it is OK to pay, and information is exchanged between the wallet and the merchant's server. The server ensures that the wallet has money and coordinates the transfer of money from the wallet to the server (encrypted, of course), and the goods are transmitted back (or shipped) to the buyer along with a receipt. Wallet software is available as a Navigator 4.0 plug-in and an Internet Explorer 4.0 ActiveX control.

17.4.1   Microsoft Wallet

The Microsoft Wallet lets customers securely store shipping and payment information. This pair of software modules is included in the latest version of Internet Explorer. A user enters address information and credit-card details into Selector sections of Microsoft Wallet. The information is then encrypted using 128-bit keys, and only the friendly name the user gives each entry is shown. For more information, see www.microsoft.com/merchant/wallet.htm.

17.4.2   CyberCash Wallet

The CyberCash Wallet enables users to make instantaneous sales transactions with any CyberCash-affiliated merchant, thus allowing users to make direct payments for goods and services over the Internet and the Web. All transactions are encrypted. Users create an account called Wallet ID which is maintained by CyberCash Servers. Wallet ID maintains the sensitive information. The CyberCash wallet will also allow the use of smart cards in electronic commerce transactions.

Users of the downloadable CyberCash Wallet can use any of three payment options to access their funds for on-line purchases:

- *CyberCoin Service* handles small transactions ($0.25 to $10.00).
- *Credit Card Services* allows use of credit cards without filling out a separate credit-card form and sending it over the Internet for each transaction.
- *Pay Now* secure electronic check service allows Internet payments to be made directly from bank account funds, again without repeatedly sending account data over the Internet.

17.4.3   Netscape Wallet

The Netscape Wallet keeps on-line payment activities under one interface. It allows users to consolidate electronic financial instruments, including credit cards, debit cards, and electronic checks. It also maintains a history of receipts and payments.

17.4.4 VeriFone vWallet

VeriFone's vWallet is a payment application that enables customers to allow SET-compliant credit-card purchases. vWallet lets consumers use their ordinary Web browsers, rather than proprietary VeriFone software, to make purchases. The system complies with the MasterCard/Visa SET protocol and includes support for electronic cash checks in addition to bankcards.

17.4.5 BlueMoney Wallet

BlueMoney Wallet from BlueMoney Inc. requires no special software on the desktop, enabling customers to make a purchase with a simple point and click. When an order is submitted to a merchant using BlueMoney software, users are temporarily moved over to a BlueMoney wallet. With first-time orders, buyers must supply credit-card numbers and a list of shipping addresses. But with subsequent purchases, users enter a secure password that will automatically pull up credit-card information saved in the wallet. Users can then point and click to select a card to pay for the merchandise.

The BlueMoney software keeps the transactions safe through a Secure Socket Layer that includes encryption. Merchants seeking continuous encryption for on-line credit-card processing are charged an additional fee per bank account as a gateway activation fee that enables the Web site to receive card authorization codes.

17.5 Digital Cash

In the early days of the Web (maybe all of two years ago), doing business on the Internet was risky. To persuade consumers to purchase products, organizations needed to develop a secure method of payment. Start-up companies created on-line systems to pass packets of digital cash securely around the Net. Customers deposited funds in an on-line account by mailing a credit-card voucher or a check to an on-line "bank," which was a start-up company itself. When customers made purchases over the Web, they used this account number and a personal identification number rather than transmitting personal details or credit-card information. Financial institutions like DigiCash and First Virtual Holdings made sure that the funds were stored safely and transferred properly. This electronic money became known as digital cash or Cybercash.

Today's wide acceptance of secure, encrypted Web connections has made MasterCard and Visa the most popular ways to pay for goods and services over the Web. Most Web browsers and Internet service providers (ISPs) support one of the major security protocols, such as SSL. For example, on Netscape's browser, if the transmission between browser and server is secure, the key icon at the lower left side of the screen is intact. Otherwise, it is split in half to signal an unsecure transmission. More elaborate methods, such as CyberCash's credit-card system, prevent the merchant from seeing the credit-card number. In addition, commerce servers from companies such as Netscape and Microsoft can verify credit-card information and process transactions in real time.

But we still have digital cash systems as an option. They may be preferred to credit-card payments for the following reasons:

- Digital cash offers anonymity and therefore protects privacy.
- Digital cash makes economic sense as a way to handle microtransactions—those transactions that are less than a dollar.
- Digital cash is still preferred by those who do not feel that the Web is safe enough for credit-card information.

Digital money can be downloaded from a participating bank into the user's personal computer as "digital coins," or a digital money account can be set up within the bank. Either the digital coins or the transactions that debit the account are transmitted to the merchant for payment. All of these transactions are encrypted for security.

17.5.1   CyberCash

CyberCash is a system of digital money from CyberCash, Inc. (www.cybercash.com). CyberCash began secure credit-card transactions in 1995, using a combination of digital signatures and public-key encryption. For a credit-card purchase, the merchant sends the customer an electronic invoice. The customer appends a credit-card number to the invoice, which is encrypted via CyberCash software and sent back to the merchant. The merchant appends its ID number to the invoice, encrypts it, and sends it to a CyberCash server for forwarding to the banking network. Debit and cash transactions are handled by preestablished accounts with a participating bank. Encrypted messages are sent to the merchant to enable the funds transfer between the participating bank and the merchant's bank.

The credit-card purchase process using CyberCash is detailed in Fig. 17-3. Most transactions take fewer than 20 seconds. CyberCash offers an electronic wallet and also, for small transactions (under $10), the CyberCoin service linked to a consumer's checking account. Both Microsoft and Netscape merchant servers work with the CyberCoin micropayment system.

Figure 17-3
Purchase Process
Using CyberCash

1. When a consumer decides to make a purchase from a Web merchant's electronic store, he or she clicks on a Pay button, launching the CyberCash digital wallet, which prompts for a credit-card choice. The transaction information is digitally signed by the consumer's wallet, encrypted, and sent to the merchant.

2. The merchant server signs and encrypts the transaction information and forwards it to the CyberCash server. Credit-card numbers are encrypted with the CyberCash public key, so merchants never handle them directly.

3. The CyberCash server decrypts and certifies the transaction data and forwards it through a private network to the merchant's bank for authorization.

4. The merchant's bank processes the charge automatically. The bank then returns an approval or denial to CyberCash, which passes the approval or denial on to the merchant. Finally, the merchant passes it on to the consumer.

17.5.2   ecash

Developed by Amsterdam-based DigiCash and the Mark Twain Bank, ecash is like having travelers checks on the Web. A user sends a check to the Mark Twain Bank, which in turn sends the user software which gives the user access to the ecash Mint. The user downloads "digital coins" to the hard drive for use when purchasing goods and services on the Internet. Payments are made by uploading ecash to a participating vendor. The transactions are encrypted to provide the necessary security for on-line transfer.

ecash (www.digicash.com) is currently used more in Europe than in the United States.

17.5.3   First Virtual

First Virtual Holdings, Inc. (www.firstvirtual.com), also offers a system of digital money. Customers establish an account with First Virtual using a credit card. When they want to purchase something on-line, they send their First Virtual ID number to the participating vendor, who in turn E-mails First Virtual and the customer for confirmation. Money is then transferred to the vendor via the private Automated Clearing House network, which processes credit-card transactions.

17.5.4 Millicent

Digital Equipment Corp. recently announced new technology for buying and selling products costing as little as a penny over the Web. Some industry watchers maintain that the lack of a model for such small Web purchases has hurt transaction processing and electronic commerce, a situation often attributed to security concerns. Digital's technology, Millicent—so named because a basic transaction only costs one-thousandth of a cent—is designed to process purchases of, for example, movie reviews, stock charts, or the temporary use of Java applets.

The technology, unlike the usual electronic payments process used by CyberCash and ecash, eliminates the need for a server by letting vendors check for frauds and double spending quickly and cheaply. Because Millicent is based on micropayments, its security requirements are much lower than those for credit-card transactions. Millicent's scrip, or electronic cash, functions like telephone cards or manufacturers' coupons. It is valid only for a single supplier. Vendors sell scrip in transactions involving real money. Users will have to establish accounts with various vendors and buy large amounts of each vendor's scrip. An alternative way of buying scrip is through brokers, which sell scrip from multiple vendors.

InfoSeek and Reuters have already adopted Millicent.

Millicent's home page is www.millicent.digital.com.

17.5.5 Internet Checks

InfoDial Inc. (www.infodial.net) has introduced what is possibly the next generation of secure Internet transaction processing for merchants: checks. Internet Checks give a consumer who doesn't want to run up charges on a credit card a secure alternative.

The consumer fills out an on-line form with both personal or business and banking information. Most of the information required is on the OCR line at the bottom of the consumer's check. The user is guided through the process of entering this information as well as personal information such as name, address, and driver's license number. Key banking information is double-entered to verify the accuracy of the entry. After the appropriate information has been entered on the transaction screen, the merchant can print a real check on a laser printer. This check can be taken to a bank and deposited.

The user of the check gives authorization for the check to be electronically signed, and the merchant indemnifies the bank, assuming full responsibility for the transaction.

17.6   Secure Transactions

How does a user know that the merchant at the other end of the transaction is trustworthy? How does a merchant know if the user really has authorization to use a given card? Encryption doesn't offer protection for these situations.

17.6.1   Secure Electronic Transactions

Recognizing that encryption wasn't enough, Visa and MasterCard, along with software and hardware vendors such as Netscape, Microsoft, IBM, VeriSign, and other major players in Web commerce, have cooperated on the Secure Electronic Transactions specification for on-line commerce. Using public-key-encrypted digital signatures, SET aims to protect on-line credit-card payments via the Internet and reduce fraud. SET complements the SSL protocol, extending security internationally.

Backed by Visa, MasterCard, and numerous major U.S. and international banks, SET creates an infrastructure for developing digital certificates, software that attests to the identity of parties in on-line credit-card transactions (see Sec. 11.2.4 for more information on digital certificates). SET uses software, digital certificates, and encryption technology to enable customers, merchants, and financial institutions to authenticate the identity of parties in a transaction in order to eliminate the risk of fraud.

As illustrated in Fig. 17-4, SET has three components:

- A consumer piece that is tied to an electronic wallet
- A merchant piece using Web commerce servers
- A gateway to a bank accepting the transactions and issuing digital certificates through a certificate authority such as VeriSign

Figure 17-4
Secure Electronic
Transactions

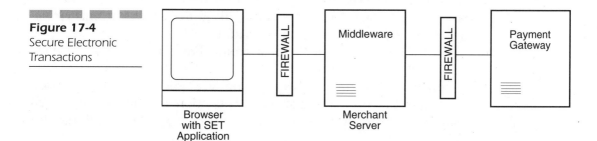

Browser
with SET
Application

FIREWALL

Middleware

Merchant
Server

FIREWALL

Payment
Gateway

Figure 17-5
How SET Works

1. A user acquires a digital certificate and digital wallet from a bank. The wallet and certificate are used to tell the bank who the user is and what credit card he or she is using.

2. The user shops at a site such as L.L. Bean. When asked how he or she wants to pay, the user selects SET.

3. L.L. Bean servers send a signal over the Internet that invokes the user's SET wallet. It pops up on the user's screen.

4. The user selects a credit card from his or her digital wallet. The digital wallet encrypts the payment information and sends it to L.L. Bean.

5. L.L. Bean verifies that the information is an SET packet and adds to the message its digital certificate, which indicates that the sender is L.L. Bean.

6. L.L. Bean encrypts the information again and passes it on to the "acquirer"—a financial institution whose business is to verify credit-card transactions.

7. The acquirer approves or denies the transaction based on credit standing and passes that information over the Internet to L.L. Bean and back to the user's wallet.

8. The transaction is complete.

The procedure used by SET (using L.L. Bean as the retailer) is detailed in Fig. 17-5. It is important to note that the merchant never actually receives the credit-card number. The number is fed directly to the bank.

SET, Version 2.0   Version 2.0 of the SET standard will add a much-needed encryption-neutral architecture that should foster the development of speedier electronic-commerce applications. Driving the need for this implementation is the server-side performance of encryption algorithms from RSA Data Security, which provides the core encryption technology used in SET 1.0. Version 2.0 is likely to implement encryption technologies from vendors other than RSA.

CyberCash changed from its own protocol to SET in the fall of 1997.

SET Mark   The companies working on SET quickly recognized that a critical aspect of how secure an SET transaction actually is, is how well managed the receiving Web site is. To assure customers that Web sites are safe for credit-card transactions, they recently announced SET Mark—a stamp of approval. A small yellow box with the SET label will be displayed on Web sites that use SET-certified commerce software. The SET Mark should assure customers that a Web site is reasonably safe for credit-card use and can be trusted as a traditional commerce channel would be.

However, the SET Mark only indicates that the software used at the site complies with SET. It does not guarantee that the site is employing other proper security measures on its own.

17.6.2 VeriFone

VeriFone (purchased by Hewlett-Packard in 1997) revolutionized secure electronic-payment delivery with its credit-card authorization products for financial institutions, businesses, and consumers around the globe. VeriFone is used by many retail stores for processing credit-card transactions. VeriFone then expanded its product vision from the merchant countertop to conducting business over the Internet and to smart-card-enabled products for the home, which Citibank endorsed.

Microsoft, Hewlett-Packard, and VeriFone are collaborating on the development of electronic-commerce systems based on the SET protocol. The companies will help merchants set up Internet storefronts, accept Internet payments, and conduct SET transactions over the Web.

Microsoft will bring its desktop presence and merchant-commerce software to the effort; HP will contribute hardware, security software, integration, and worldwide support; and VeriFone will contribute Internet-payment technology, its relationships with financial institutions, and its SET expertise.

In addition, Microsoft and VeriFone also plan to jointly develop an SET-compliant payment module for Microsoft Wallet, a PC-resident application that communicates consumer-purchasing information to merchants and financial institutions.

17.7 Storefront Infrastructure

Launching a successful on-line storefront takes more than artful Web pages. An organization needs to take orders, secure transactions, track the inventory, and keep the site up-to-date.

The first requirement is a Web server that supports secure transactions.

As is usually true of platform decisions, this decision should be made after all the software needs have been determined.

Shoppers will need a way to search for items they want to purchase. A payment system that either accepts credit cards over the Internet, uses a Net-based payment scheme such as CyberCash, or is SET-compliant (for credit-card transactions) must be in place.

Begin by looking at database servers and their associated Web-based tools. The core task of any Web storefront is manipulating, storing, and updating data—not only on the items for sale but on those in inventory and on order, as well as customer information. If data is to be exported to a Web commerce software, existing accounting, inventory, and product-catalog data will need to be generated in comma-separated files. This is the universal data-interchange format—for now.

If this information is already maintained in a database, then run the Web storefront using that database software. If not, consider Microsoft's SQL Server or Informix's or Oracle's Web and database servers, all of which run on NT. These are common platforms for Web-based storefronts, and a wide selection of Web-based tools work with them. SQL Server is the best choice if an organization is planning on using other Microsoft-based tools, such as Microsoft Commerce Server or Microsoft's Web Server. Use of Oracle requires expertise in Oracle's PL/SQL language. Informix is a fine choice if the organization doesn't have any PL/SQL expertise and isn't married to Microsoft.

After choosing a database, choose a tool to manipulate the data into Web format. There are many, but most fall into one of three categories:

- Products such as Allaire's Cold Fusion that add proprietary HTML-like commands to Web pages, so that results from users' queries can be built on the fly. These products are good for developing custom Web-based forms, such as those that let visitors search an entire on-line catalog for a particular feature.

- Application-development environments, such as Bluestone's Sapphire, that can be used to publish parts of a database on the Web. Developers design the forms, queries, and data structures, and these products will create CGI scripts automatically. Use these if the intent is to Web-ify already developed client/server databases.

- Products such as Oracle's WebServer that create queries and Web forms using their own SQL-like commands.

Depending on the organization's database expertise and its goals for the electronic endeavor, a mix of product types may be necessary.

17.8 Success Stories

Retailers have mixed approaches to the use of browsers and Internet strategies. Some are investing in consolidating their internal networks using i-net technologies. Others—especially those that already have catalog sales—are going on-line to conduct electronic transactions while their internal networks take a back seat.

Through all the noise about advertising-driven Web business models and the naysayers promising that selling products on-line would never work, the reality is that if it's done right, it can be great for business.

Dell Computer Corp. Dell Computer Corp. has invested millions in a worldwide digital storefront and is now reaping the profits from its on-line gold mine. According to Dell officials, the company's Web site generates $1 million per day in system orders—orders that have been increasing by 30 percent each month. And although only half of its on-line customers fit the profile of Dell's traditional corporate clients, the company claims to have no problem selling servers costing as much as $30,000 over the Web.

Dell, expecting E-commerce competition from the big players such as Compaq, IBM, and Hewlett-Packard, is turning its focus to offering customized Web pages for larger corporate customers as some corporate customers are softening their resistance to making major purchases over the Internet.

Under this initiative, called Premiere Pages, customers will be fed information relevant only to their individual business, such as predetermined system configurations and pricing discounts. These company-specific pages also list contact information, including pager numbers, for each high-volume customer's assigned account representatives and dedicated technical support staff. Customers also will be able to check the status of orders and place service requests over the Web. Premiere Pages allows Dell to configure and ship systems for a customer to any number of sites anywhere in the country.

To improve efficiency, Dell has pulled its component suppliers into its on-line inventory management system, which will enable suppliers to automatically send parts when Dell runs low.

Cisco Systems The networking vendor Cisco Systems' Cisco Connection Online (www.cisco.com) pulled in more than $2 billion since October 1996. Estimates for 1998 are $3 billion in sales. If Cisco's estimates are correct, on-line transactions will amount to about 40 percent of its total sales.

Cisco Connection Online (CCO), the umbrella name for all E-commerce functions on Cisco's Web site, has 50,000 registered users. Cisco conducts more than 70 percent of its customer support through its Web site and handles 15 percent of product orders over the Internet. CCO gives users direct, real-time access to information and services such as software upgrades, technical assistance, order status, seminar registration, documentation, and training. Cisco says that doing business on the Net saves the company $250 million a year in customer-support expenses by moving support functions to the Web—$100 million of the savings was in printing costs alone.

Recreational Equipment Inc.   Recreational Equipment Inc. (REI) has a virtual store with over 2000 products to choose from. Customers can place an on-line order and pay for it with a credit card over the Web. The REI On Line store, in place since late 1996, averages 5 million hits a month. REI reports that on-line transactions are 60 percent above its expected target.

17.9   Web-Spawned Businesses

Many organizations are looking to the Web as another link to their customers. Many are starting businesses that can be successful only on the Web.

Jerry Kaplan   Jerry Kaplan holds auctions of overstock, end-of-life, and refurbished PCs over the Web (www.onsale.com). With up to 25,000 visitors per day, the company Onsale pulls in roughly $1 million a week in sales.

Peter Ellis   Car salesman Peter Ellis was nearly wiped out by the California recession of the early 1990s—he had to close or sell off 16 dealerships. By 1995 he was back in the car business on the Web. Auto-By-Tel makes money by selling sales leads to auto dealers across the country. For a monthly subscription fee, dealers get the names of Web surfers who have visited www.autobytel.com for manufacturers' prices. The nearly 1600 dealers love the excellent sales leads. The Web surfers are happy to gather information without having to deal with a car salesperson.

Happy Puppy   In 1995 Happy Puppy, the Internet's top game site, was started as a means to market games by putting demo versions on-line. Today visitors to www.happypuppy.com can sample the hottest games on the market. The site makes money from selling software as well as from advertising.

E*Trade   Fast-growing E*Trade Group Inc. (www.etrade.com), an Internet stock trading service, expects revenue for this fiscal year to reach $126 million and profits of nearly $12 million. It has, on average, 550 new customers a day and handles about 75,000 stock transactions a day, and the portfolios it represents control $3.2 billion in assets. All at $14.95 per market-listed trade—a quarter of the cost per trade through a discount broker. See www.etrade.com for more information.

On-line stock trading has become one of the few industries that are making profits from Internet-based applications. During mid-1997 a typical day involved 95,000 stock trades. On such a day, brokerages made close to $2 million in commissions and investors saved at least $7 million by buying on-line rather than paying the higher commissions of traditional brokers. A recent survey by Forrestor Research found that 30 percent of on-line stock brokerages were profitable and another 20 percent were breaking even.

Amazon.com   Founded in 1994, the Seattle company Amazon.com Inc. uses on-line connections to major book distributors to offer 2.5 million book titles. Amazon.com's success so far is built on customer focus, which means a well-designed, easily navigable Web site, located at www.amazon.com, on the front end, and fast, efficient fulfillment and delivery on the back end.

Amazon.com filed an initial public offering that put the company's value at roughly $300 million in April 1997. It originally planned to sell 2.9 million shares at $13 a share but increased the number of shares to 3 million in hopes that the company would raise $42 million. In the end, Amazon.com raised $54 million. It is important to note that Amazon.com has still not made a profit, however.

Amazon.com gets more than 80,000 visitors daily. That's about 29 million visits a year before any growth. The big names in bookstores sat up and took notice. Barnes & Noble opened its Web site in mid-1997, and Borders is in the process of bringing its site on-line.

17.10   Keys to Success

The best advice from those that have tried and succeeded in electronic retailing is for the organization to know why it wants to sell over the Net.

Organizations need to remember that the demographics of those who surf the Net do not reflect those of society as a whole. In addition, not everyone who surfs the Net thinks that it is secure enough for transactions. Customer service becomes the overriding issue. An electronic storefront has to be easy to use, provide the customers with what they want, and be secure.

17.10.1   Control Issues

Buyers are using the Web because it puts them in control of the purchasing process. Cisco and Dell are letting users price, configure, and buy products on-line. Every system specification can be selected by the user, often via handy dropdown menus. Scripts built into the system can automatically alert the customer if there's a compatibility conflict among chosen components or if a given configuration is unavailable. In a well-constructed electronic storefront, the buyer can control everything except the product's price.

Web buyers are the most price-sensitive buyers on the market; they know what the product they are looking for costs, and they refuse to pay a dime more. With a check of the competition's latest price only a click or two away, the Web is becoming a battlefield in a high-tech price war.

17.10.2   A Way to Do Business

Customer service does not disappear because there is no face-to-face, or voice-to-voice, contact. Web commerce is a way to do business—just having the storefront will not make the sale; customer service will.

For example, Dell's Web site is also its primary tool for delivering technical support to users. The company has 30 people constantly updating some 35,000 support-related HTML pages. The phone-based tech-support reps read from the same pages a user can find at www.dell.com. In the future, the company intends to use the Web as the launch pad for an array of self-diagnostic services.

Enhancing the quality, not just the quantity, of Web transactions can affect the nature of sales as well. According to Dell, the amount of the average on-line order is significantly higher than that of the average purchase by phone. Dell's Web site even makes money on users that don't buy on-line. The company's research shows that customers calling Dell's 800 number are twice as likely to buy a system if they have previously checked it out on the Internet.

17.10.3 Serve the Interests of the Consumer

Use of the Web is reengineering at its best. Successful organizations have focused on the interests of their customers, not on their marketing department. Amazon.com, Dell, Onsale, and many others have developed a business strategy that truly puts the consumer in the driver's seat. Consumers are flocking to those sites. And consumers are buying at those sites.

18

Software for On-line Shopping

Commerce is about buying and selling. The process itself can be broken into four stages:

- Before the sale
- Making the sale
- Getting the goods
- After the sale

Figure 18-1 illustrates how these are done traditionally and how the Web can be used to enhance the way these stages function.

Electronic retailing software addresses these four areas. Electronic storefronts focus on the before the sale and making the sale activities—the "front end." Commerce servers handle the "back end"—getting the goods delivered and maintaining after-sale activities.

18.1 Storefront Software

Organizations can build their electronic storefronts with their existing Web servers or use E-commerce suites that include both the storefront features and a server optimized for E-commerce.

Figure 18-1

Impact of Web Technology on Stages of Commerce

	Process	Web-Based Enhancements
Before the sale	Sellers need to attract buyers somehow. Buyers have to be able to find sellers. Buyers want to browse before making a decision.	Electronic marketing can take the form of Web pages, product catalogs, and banner ads on other sites such as AOL or one of the search engines. Buyers should be able to type in a search term and see what products meet that criteria.
Making the sale	Sellers must provide familiar methods for making purchases. Sellers may want to make discounts or special offers to sweeten the sale. Buyers want to be able to choose payment methods and shipping options.	Web software uses shopping carts as one way to make customers feel comfortable. There should be automatic tax and shipping calculations, and the cost of the order should be totaled automatically. Payment information has to be secure.
Getting the goods	After a buyer has paid for a purchase, the seller must deliver the goods promptly.	Internal systems have to be in place to ship hard goods that are ordered, as customers expect to see them as quickly as if they were ordered by phone.
After the sale	The all-important customer service. If customers are treated right and like the product, they will come back again and again.	If a customer calls in with a question about an order—or, better yet, uses the Web to research the status of an order—internal information must be available. More importantly, it shouldn't make a difference whether the purchase was made over the Web or over the phone. Buyers can also become part of the process. They can participate in chat rooms, or give endorsements that could then appear on the seller's Web pages.

All-purpose storefront packages should support Secure Socket Layer (SSL) and Secure Hypertext Transfer Protocol (S-HTTP), with plans to support the Secure Electronic Transactions (SET) standard when it is available. Although SSL and S-HTTP can securely transmit information such as credit-card numbers, most credit-card issuers urge their cardholders to do on-line business only when using SET.

CyberCash and VeriFone offer secure alternative payment options that verify that a customer is a cardholder and send credit authorization to the merchant. Most storefront products support CyberCash and VeriFone as well as SSL and S-HTTP.

Some companies are also setting up shipping and tax-calculation tie-in products to assist in automating the transaction. For example, Taxware International offers sales-tax verification via a database of tax jurisdictions keyed by ZIP code.

For convenience, most storefront software integrates a Web server into the package—you can't have a Web store without a Web server!

Most storefront software includes predesigned storefronts for the Web merchant to start from, as well as templates for such things as shopping carts, database services, and tracking customer visits. Some knowledge of HTML (Hypertext Markup Language) is required—sophisticated customization requires more than a basic knowledge of HTML coding and design practices.

Most Web storefronts are running Windows NT 4.0 and support Open Database Connectivity (ODBC) and database servers.

18.1.1   Cat@log 1.0

The Vision Factory's Cat@log 1.0 works with an existing Internet server and an existing catalog database to dynamically generate Web pages. While it's a little more difficult to use than all-in-one builders like iCat's Electronic Commerce Suite, it offers more store-building power at a comparable price.

Cat@log has four major components: Builder, Manager/Online Publisher, Commerce, and Customer Statistics. Cat@log Builder is the module used to construct the site. It generates both static and dynamic Web pages, including the forms and query pages. Builder lets the creator quickly navigate and organize the site using a visual hierarchy, and it's flexible enough that pages can be designed any way the creator wants.

Builder uses both a third-party graphical Web-page publisher and a text editor to tweak the HTML. While this is a core strength, since it provides a lot of flexibility for designing pages, it also means that Cat@log requires a greater level of Web-building expertise than some of its more "pushbutton" competitors.

Cat@log **Manager/Online Publisher** arbitrates queries between the Web catalog and the product database, generating search results in the form of catalog pages seen by on-line customers. Manager runs in the background, called by the Web pages whenever a dynamic page needs to be generated.

The Cat@log **Commerce** module integrates Manager with secure third-party payment authorization services like VeriFone. The package's most useful feature is the **Customer Statistics** module: It provides statistics on store traffic, customer shopping profiles, and the like—necessary information for anyone who wants to build an effective storefront.

All four components of Cat@log are sold separately. While one could theoretically deliver a site using only Builder and Manager, in practice all four components are required.

Cat@log lacks complete remote-management features, so most of the serious site maintenance needs to take place wherever the server is physically located. Also, there are no native inventory-management features, which would tell customers that a certain product is out of stock before they put it in their shopping baskets.

For more information, see www.thevisionfactory.com.

18.1.2   Cashier 3.0

Cashier 3.0 from InfoDial (www.infodial.net), in addition to Internet Checking, has the following features:

■ Support for full database integration allows the merchant's product line and pricing to be stored on-line and fully integrated with transaction processing. A product search engine with output controls and real-time price/information/availability controls is also available.

■ QuicKart single-form purchase ability allows for instant purchasing, and Shopping Cart Integration automatically figures subtotals and totals.

■ Credit-card authorization is accomplished in real time, with support for virtually all processing networks. Instant real-time transaction results are provided. Complete security is provided, since all transactions take place with SSL encryption.

■ Real-time management and control is offered, as well as E-mail notification of orders, without exposing transaction information.

Cashier 3.0 supports all SSL-compatible browsers. InfoDial customers simply pay a normal monthly fee for Web-site hosting. There is no percentage-of-sales charge or any other type of override on the normal monthly fee.

18.1.3   Intershop Online 1.1.4

Intershop Online comes with ODBC support as well as Sybase's SQL database Server 11. Intershop Online provides back-office functions and has built-in manager templates for administering catalog/storefront, product, order, inventory, supplier, and customer functions.

Intershop Online (www.intershop.com) can be used to create storefronts in four languages: English, French, German, and Norwegian. The text for the products being sold has to be rewritten in each language, but the database schema stays the same.

The product also has sales analysis and reporting functions and supports remote administration via a Web browser.

18.2   E-Commerce Suites

The decision on the software suite follows the database server decision. Typically, such a suite contains four elements: a catalog or database to which you add items to sell, a storefront designer to arrange them and to add graphics, an ordering and inventory system, and a shopping cart system that allows customers (and the seller) to keep track of purchases as they move from page to page and to pay for their goods. These elements can range from a series of common gateway interface (CGI) scripts to more sophisticated applications, depending on the suite.

Unlike productivity suites such as Microsoft Office and Corel Office Professional, E-commerce suites are more like toolkits. There is a great deal of integration, setup, and programming required to open a storefront. For example, ZIP codes need to be matched with regional sales taxes, and prices need to be adjusted to reflect promotions.

Why use a suite? The suites are a good place to learn by example. Each comes with at least one sample storefront that can be modified and used to start a store. The suites also come with shopping cart systems, making it easier to develop a way to track what shoppers choose.

Third parties are providing APIs (application programming interfaces) to their products. TanData Corp.'s Progistics.Merchant automates the shipping and handling of goods. Internet Tax System from TaxWare International Inc. enables a site to build in automated tax calculations. These applications are available separately.

18.2.1   Microsoft Commerce Server

Microsoft's first product in this arena was Merchant Server, which was first available in December of 1996. It has since been replaced by Commerce Server, which is included in Site Server Enterprise. Commerce Server runs only on the NT operating system and requires at least Version 2.0 of Microsoft's Web server, Internet Information Server (IIS).

Commerce Server comes with components for creating shopping carts and payment systems, and storefront designers to help manage the store, set up the database, process sales, and handle accounting. Commerce Server supplies a list of items for sale and buttons that let users search the directory, view their shopping basket, and check out. If the organization already has product and financial databases and can link to them with ODBC, this information will not need to be duplicated inside Commerce Server.

Commerce Server comes with three environments: development, staging, and production. Changes are posted to the development environment, where Commerce Server debugs HTML pages. Once the results are satisfactory, the data is migrated to the production server, where Commerce Server removes any debugging code. All the dynamic page-generation logic is written as Active Server Pages driven by VBScript or JavaScript. (Merchant Server used a proprietary generation system.)

Tools are also provided in the database server and Registry Editor for migrating data from development to production. The first step is to copy the database schema and associated data tables, any supporting files (such as templates and graphic images), and Registry keys.

To set up a new store, the Administrator control panel is used to copy one of the four starter businesses—a sporting goods store, a bookstore, a clock store, or a coffee shop. Each has different item groupings and configurations, but the elements from the stores can't be mixed and matched. For example, the sporting goods store has the best merchandising techniques, with support for sales, promotions, and cross-selling. The simplest model, the clock store, has only one product type and lacks search and registration functions.

The second step is to run the ODBC control panel and create a new data-source name to link to. Using a text editor, SQL script files are modified to replace the name of the starter store with the new name. These scripts are then run by Microsoft's SQL Enterprise Manager, the Registry entries are modified for the new store, and the Commerce Server is restarted from its control panel.

There are entries to enable cookies (text files of information sent back and forth between a browser and a Web server—see Sec. 6.10 for more information on cookies) to store user identities, to allow SQL Server to start before the Merchant service starts, and to remotely administer Commerce Server over the Internet.

Commerce Server comes with the VeriFone payment system, which is a plus if the organization already has a merchant bank account set up to accept credit-card payments from non-Internet customers.

Developers can use the components as is, write their own, or purchase additional components from third parties.

Besides Commerce Server, Site Server includes

- Microsoft Personalization System, a toolset for delivering dynamic, customized Web content to users
- Content Replication System (CRS), a tool to synchronize content on various servers
- Site Analyst (formerly WebMapper from NetCarta), a tool used to gather information about a site's content and identify problems such as broken links
- Usage Analyst (formerly Market Focus from Interse), a tool used to analyze Internet log files

If the organization is already NT-based, has its inventory in an ODBC data source, and is a VeriFone customer, Commerce Server is an ideal choice. However, if the organization is weak on database technology, it may not be the ideal choice.

18.2.2   iCat Electronic Commerce Suite 3.0

The heart of iCat (www.icat.com) is its catalog of store items and the tools for managing them. It also has software for designing Web storefront pages and tracking purchases. As with Merchant, any changes are posted to a staging database and then copied to the live production database.

Setting up a store with iCat isn't difficult. First, the database is created, either using the included copy of Sybase SQL Anywhere or connecting to an organization's ODBC source. The promotional prices, discounts, and upsells are entered at this point. (An upsell is suggesting an item to a customer based on something already purchased. For example, CDs to go with that new CD player?)

Next, templates are designed to control the look and feel of the pages. Third, the location of all HTML pages, databases, and other files being used is specified. Finally, sales options are set—for example, specifying sales tax rates for particular locations. Many of these options are already part of the sample hardware store database.

The sample store is well designed and because of this, it's easy to see how various screens relate. For example, the sales tax screens match a sales tax rate with a ZIP code. Both pieces of information (the ZIP code and the sales tax rate) still must be entered, but at least the data structures are specified.

iCat comes in two versions: standard and professional. A cottage industry of third parties has grown up to supply iCat with add-ons, including payment-authorization software and sales tax calculators. These add-ons work only with the professional version. iCat's Commerce Suite also has a Web-based catalog editor that lets you add items to your store and change prices.

For those who are new to databases or the Net, iCat is a good choice. It has excellent tools for beginners with little HTML expertise and offers easy-to-set-up shopping cart and accounting systems. If the organization already uses computer-based accounting or inventory tools, it may have some trouble integrating the storefront with existing systems.

18.2.3   IBM Net.Commerce 1.0

IBM's Net.Commerce (www.internet.ibm.com/net.commerce) comes with everything needed to set up shop: a Web server (IBM's Internet Connection Secure Server), a database server (IBM's DB2), and E-commerce pieces, including shopping cart and credit-card verification applications.

Net.Commerce requires a VeriSign SSL certificate to build the store as well as to provide encrypted browsing for customers. After the secure key is activated, the DB2 and Net.Commerce servers can be installed and then rebooted. Next, the DB2 server, the Web server, and the Net.Commerce server are started, using NT's Services applet in the Control Panel, and the configuration process is begun.

Net.Commerce includes four Web forms that contain the controls needed to get the storefront up and running:

- *System Configuration,* which has basic Web-server details on directory locations
- *Access Control,* with user identities and password information
- *Server Control,* which can start and stop the Web and database servers
- *Database Management,* which tells the software the database name and TCP port number

Net.Commerce is an ideal choice for businesses that already use IBM's DB2 for their inventory system and for merchants who want to build their storefronts around VeriSign. But if DB2 isn't one of the onsite databases, using Net.Commerce is probably not the way to start.

18.2.4   Netscape Merchant System

Besides its popular Internet servers, Netscape also offers the Netscape Merchant System for complete retail management services. Web merchants can display thousands of products in a catalog, handle hundreds of simultaneous transactions, and integrate with legacy database systems for inventory, orders, and fulfillment. A staging server is used to develop the Web store, while product offerings migrate to the Merchant Server and a separate Transaction Server handles orders.

Merchant System integrates merchandising capabilities, including product search, promotional discount support, flexible pricing, dynamic displays, and multimedia integration.

First Data processes payments, and the system integrates with Taxware's sales tax computation engine and with Netscape's other high-end product, Netscape Publishing System, which lets publishers create subscription-based publications. Merchant System is available for Sun Solaris and Silicon Graphics Irix, with other Unix versions expected in the spring of 1998.

18.2.5   Open Market OM-Transact

Open Market Inc. (www.openmarket.com) is in the business of supplying industrial-strength commerce software, with an emphasis on back-office infrastructure, for high-volume Web merchants—and has been doing so longer than other vendors in this arena.

OM-Transact provides a complete back-office infrastructure for secure Internet commerce and supports secure payment, order management, transaction processing, and customer service for high-volume transaction environments. As a backbone Internet commerce offering, it works in conjunction with Web-site and catalog-development tools such as those available from Cadis, Saqqara, and Bluestone, all of which are Open Market software partners, as is the iCat catalog product. OM-Transact runs on major Unix platforms such as those from Sun, Silicon Graphics, Stratus, and Hewlett-Packard.

Its architecture allows a Web site to separate access control and commerce functionality completely from other content and services. This allows for increased security as well as allowing Web sites to outsource this functionality and Internet service providers (ISPs) to offer it as a service to Web-site developers.

Open Market's Web-based electronic commerce tools are based on its Open Market Secure WebServer. OM-Transact handles transactions. Merchant Solution lets merchants run their entire business on the Web. Users can subscribe to Open Market's Integrated Commerce Service, which features order taking and credit-card payment authorization, sales tax calculation, digital receipt generation, and computation of shipping fees. S-HTTP is used to secure transactions. OM-Axcess enhances S-HTTP security by issuing cryptographic objects called access tickets, which uniquely identify each user and grant permission to access company information. The access ticket method gives administrators increased control over access to resources, and it measures and reports access to data across numerous vendors' Web servers.

Open Market recently acquired Folio Corp. and uses its products as a front end to OM-Transact and its SecureLink application. This combination allows Web content companies to deliver information on-line and charge subscription fees. Folio's Site Director Web-site software has a metering API that allows site managers to control access to documents and determine who has been looking at them.

18.2.6   Up-and-Coming Suites

Oracle and Lotus Development Corp. have recently released products that bear watching. In addition, Actra Business Systems, a joint venture of Netscape and General Electric Information Services, now totally funded by Netscape, began to release suites of servers as part of its CrossCommerce initiative in late 1997. The first product, ECXpert, focuses on EDI (electronic data interchange). By early 1998, Actra expects to have OrderXpert Seller, OrderXpert Buyer, MerchantXpert, and PublishingXpert on the market; these products are aimed at electronic storefronts.

Internet Commerce Server from Oracle   Oracle's Internet Commerce Server (ICS), formerly code-named Apollo, combines a number of Oracle database and Web applications with third-party payment offerings. ICS is a cartridge that fits into the Oracle Web Applications Server 3.0 Advanced Edition (www.oracle.com). With the package, users receive an Oracle 7 database, the ConText search server, VeriFone's Virtual Point of Sale software, and CyberCash's micropayment software.

The new server also comes with Store Manager, a Java-based management application that includes customizable HTML templates and authoring tools for building and managing an on-line storefront. Store Manager also provides a staging system for testing the storefront before launching it.

Java Shopping Cart is an applet that stores buyers' purchases as they surf E-commerce Web sites. The applet also totals the bill for the products and sends the information to the merchant server to complete the transaction.

Java Wallet manages a user's credit cards, electronic cash, digital certificates, and other payment information.

The server software also includes support for automatic generation of catalogs out of the database and batch loading of information into back-end financial applications, as well as shopping cart capabilities and the ability to check on the shipping status of goods purchased at a site.

Domino.Merchant from Lotus Lotus (www.lotus.com) offers Domino.Merchant, an application for the Domino server, and has plans for further integration with parent IBM's E-commerce offerings. Domino.-Merchant offers product catalogs, shopping cart capabilities, order processing, tax calculations, and various payment features.

With Domino.Merchant, an organization can ask all its suppliers to post their catalog and product documentation updates on a Web server outside the company's firewall. Once a day, the company's Webmaster replicates any changes to the Domino databases inside the firewall.

In late 1997, Lotus upgraded the application for IBM's Net.Commerce on inventory management, relational DBMS access, and Secure Electronic Transactions using Visa or MasterCard. The application also will be able to support Secure Socket Layer 3.0 encryption, which was included in Domino 4.6, released in late 1997.

18.3   Suite Alternatives

Suites aren't for everyone, though. Don't buy one if you have CGI and PERL programming expertise or plan to use a consultant who has these skills. The "homegrown" PERL solution, developed for a shopping cart system such as the one at www.egrafx.com/minishop, provides many of the same functions for about the same cost. If you're already using a combination Web-and-database product such as Cold Fusion or Sapphire, you probably can duplicate many of a suite's functions at a much lower cost.

If you can't afford a storefront suite, consider single-function tools. You could use a shopping cart system from Mercantec, a search/index tool from Saqqara, and an accounting tool from Inex and still spend much less than you would for the least expensive suite.

18.4   Web Servers

Once the software issues have been addressed, the Web-server decision is easy: Pick the one that the chosen database and software tools support. The server should support at least one, and preferably both, of the two most common secure protocols: S-HTTP and SSL. S-HTTP encrypts the Web-based traffic between client and server on a page-by-page basis. SSL encrypts more of the protocol stack and works on a connection-by-connection basis. Neither is bulletproof, but when they type in a credit-card number and sent it over the Net, shoppers like knowing that the data is encrypted. Some servers, such as IBM's Secure Internet Connection Server, support only one of the systems; others, such as O'Reilly's WebSite Pro, support both.

The most common secure Windows NT Web servers are those from Microsoft, Netscape, O'Reilly, America Online, and Oracle. There are plenty of others to choose from; try www.webcompare.com for a long list. Once the Web-server software is in place, the final step is to obtain a certificate from VeriSign; this is necessary if the server runs SSL, because SSL operates on a connection-to-connection basis. The Web-server software is used to create a certificate request. The information is printed and faxed back to VeriSign.

See Sec. 9.3 for more information on Web servers.

18.5   eTrust

While digital certificates and other advances are paving the way for secure Internet transactions, businesses and consumers need reassurance that they can trust the technology. Enter eTrust from the Electronic Frontier Foundation and CommerceNet, an effort to increase the amount of trust between buyers and sellers doing business in the electronic marketplace. eTrust (www.etrust.com) is a licensing and accreditation system for on-line merchants that will give consumers an assurance of privacy and security during on-line transactions. The eTrust logo system will be administered by the accounting firms KPMG Peat Marwick LLP and Coopers & Lybrand LLP, which will audit companies and monitor their on-line practices concerning the disclosure of user information. Companies that meet accreditation standards can license the eTrust logo and display it on their Web pages.

Under the eTrust program, there will be three different logos that will permit merchants to collect and distribute varying degrees of information from consumers who visit their Web sites.

eTrust will work in conjunction with existing electronic commerce standards, such as SET. But efforts like SET will not matter unless consumers trust the concept of electronic commerce as a channel, the eTrust partners say. Hence, a seal of approval to businesses that adopt certain standards ensuring consumer privacy and security should give consumers the confidence they need to make electronic purchases.

18.6   Specialty Web Development Tools

Some organizations are creating development tools targeted at particular industries. Microsoft is the first to have seen such a need and filled it.

18.6.1   Microsoft Internet Finance Server Toolkit

Microsoft recently released a Windows NT Server–based Web-site development package called Microsoft Internet Finance Server Toolkit (MIFST), formerly code-named Marble, designed to help financial institutions set up and run secure E-commerce applications. MIFST has scripting tools to enable communication for bill payment and other services, an extensible server component, and a gateway connection to existing financial applications. Also included is a server application component that supports 128-bit encryption and works with most firewall systems.

A U.S. Department of Commerce license obtained by Microsoft in 1997 allows the company to issue digital certificates to international banks, enabling them to use 128-bit encryption without key escrow. Key escrow programs store coding and decoding keys with a third party in the event that the encrypted data needs to be accessed by someone other than the user. Key escrow programs are explained in more detail in Sec. 11.2.

MIFST complies with an existing standard interface, the Open Financial Exchange (OFX) specification, enabling it to work with popular financial applications such as Intuit's Quicken and Microsoft's Investor and Money. OFX is a proposed standard from Microsoft for home banking on the Internet. It complements Microsoft's Open Financial Connectivity (OFC) specification, which would standardize integration with banks' back ends.

MIFST's scripting tools can be used by developers to link legacy systems to commerce service providers such as Checkfree Corp.'s bill payment service. MIFST includes customized HTML templates and can be customized with Java, ActiveX, Visual Basic, or Visual C++.

MIFST was released at the end of 1997. CoreStates Bank, N.A.; Paine Webber Inc.; and Wells Fargo were among the first companies to implement on-line services solutions based on MIFST. The MIFST toolkit provided these companies a means of delivering support to customers using personal financial management software such as Microsoft Money or Intuit Quicken through their own Web sites or Web clients such as Microsoft Investor.

19

Web-based Electronic Data Interchange

The electronic data interchange (EDI) technology is nearly 20 years old. EDI grew out of an organization's desire to complete business transactions in days, as opposed to weeks or months as was the case with paper processing, manual data entry, and postal mail. In today's fast-paced economy, organizations are looking to measure this in hours rather than days.

From its current basis in closed, proprietary wide area networks and value-added networks, EDI technology is quickly moving toward the open, free space of the Internet. While the opportunities are practically endless, the shift is not without its hurdles.

19.1   Definition of EDI

EDI is a collection of standard methods for codifying certain business documents for electronic transmission. Vendors and their customers use EDI to link their computing infrastructures without worrying about the differences in their respective systems. Vendors, distributors, and suppliers that frequently do business together can save time and money with EDI by eliminating or reducing paper transactions.

The electronic documents—purchase orders, shipping documents, invoices, invoice payments, etc.—are transmitted between an enterprise and its "trading partners." A trading partner, in EDI terms, is a supplier, customer, subsidiary, or any other organization with which an enterprise conducts business.

EDI software translates fixed-field or "flat" files that are extracted from applications into a standard format and hands off the translated data to communications software for transmission. The received data is then translated by the receiving computer and formatted for its applications.

The formats used to convert the documents into EDI data—and back again, of course—are defined by international standards bodies and by specific industry bodies. Third parties provide EDI services that enable organizations with different equipment to connect.

In practice, EDI is difficult to implement and manage. As a result, smaller companies traditionally have not been able to use EDI and have turned to an EDI clearinghouse, a company that provides EDI capabilities for a fee.

19.2   Implementation

EDI was developed to improve operational efficiency. It forced trading partners to work together to improve efficiencies on both sides of the connection.

19.2.1   Traditional Implementation

A traditional architecture is illustrated in Fig. 19-1. Traditional EDI systems were built around private wide area links and mainframe or midrange computers. Companies set up dedicated, proprietary links with their trading partners. Though based on legacy technology, many of these systems are still in use today.

Companies conducting EDI must buy software to translate their busi-

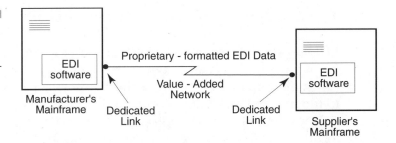

Figure 19-1
Traditional EDI Architecture

ness information to and from the EDI format and integrate EDI with their internal business applications. Prices range from $5000 for PC-based systems to $250,000 for mainframe applications. Companies must also sign up with value-added network (VAN) providers, which charge a monthly fee as well as per-transaction fees, which can add up quickly.

The closed, proprietary connections used for EDI allowed secure transmissions between trading partners. However, the complexity of traditional EDI systems and the high cost of implementation (which could reach millions) made them economically viable for only a few corporations. In addition, because of their complexity, these systems often took years to design and implement, so that they were likely to be outdated by the time they were rolled out. In order to implement an EDI system, companies have needed the necessary computing power, the staff to support it, and connectivity experts to help them connect to their various trading partners. Because it is expensive, complicated, and proprietary, EDI has long been considered an elitist technology and has remained out of reach for smaller businesses.

With traditional EDI, the communications architecture is a big part of the expense. Proprietary VAN providers such as General Electric Information Services (GEIS) and Sterling Commerce assume responsibility for transmitting, controlling, logging, and archiving all messages through a central electronic clearinghouse. The charges for setups and communications exchanges can easily reach tens of thousands of dollars a month. As a result, EDI has seen a very low penetration level. While most of the *Fortune* 500 companies are equipped with EDI, only 6 percent of the remaining 10 million U.S. companies are.

EDI service bureaus popped up to act as intermediaries between EDI users and small suppliers. The service bureaus take EDI notices from a hub's value-added network mailbox, send that EDI information to the specified supplier, translate that supplier's documents—such as invoices or shipping notices—into EDI format, and send the result back to the EDI user.

19.2.2 Internet Implementation

EDI never gained widespread acceptance, but it is set to make a comeback by way of the Internet. Network managers in small and medium-sized companies can expect to have at their disposal a new generation of easy-to-use Internet-based EDI services and interoperable software.

Until now, EDI, which has operated on proprietary networks and platforms for the past 20 years, has attracted only about 100,000 users worldwide. And though EDI software is available from a few software developers, these products are not interoperable, which forces trading partners to use the same platforms.

But Internet-based services and interoperable software promise to make EDI more widely available to those interested in conducting this form of E-commerce over the Internet. The combination of low-cost services and interoperable TCP/IP-based applications allows organizations of all sizes to benefit from EDI technology. The Internet provides large-scale connectivity without any special networking infrastructure or software, allowing EDI trading partners to conduct business with the widest possible range of companies. The architecture for these extranets (an intranet that allows select outsiders in) is illustrated in Fig. 19-2.

Figure 19-2
Web-based EDI

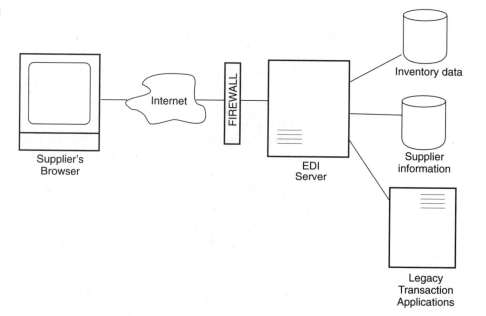

The Internet has leveled the playing field. Any business that can afford a PC, a modem, and the monthly Internet service provider subscription fee can participate in EDI commerce. Smaller businesses with fewer resources can compete head-to-head with their larger counterparts. Consider the success of Amazon.com as it competes against the likes of Barnes & Noble.

Business-to-business E-commerce involves applying Internet technology to existing business processes, and then forging critical links to legacy databases and order-entry and customer-service applications. It isn't about huge new revenue channels; it's about savings and customer loyalty.

Internet EDI systems are easier to implement and adapt readily to technological changes. The Internet is widely accessible to potential suppliers and consumers. Large organizations with traditional EDI systems are also supporting EDI on the Net just to stay competitive.

Today a small company that wishes to interact with a larger company can do so by installing EDI software and communicating over the Internet or can turn to an EDI boutique, which is an outside vendor that provides EDI capabilities on the Internet for a fee, much like the original EDI service bureaus.

Previously, smaller suppliers became involved with EDI only when they were forced to do so because they wished to trade with much larger firms. For many, the cost caused them to decide not to do business with those larger firms. Now the Net provides the communications capabilities of a VAN at a far lower price. The Internet Engineering Task Force is actively pursuing standards for encapsulating EDI documents within the Secure/Multipurpose Internet Mail Extensions (S/MIME) protocol. Web pages can be set up to contain EDI documents while concealing their complexity from the user.

EDI systems were driven by dominant industry players. Today, smaller companies within specific industry sectors are joining forces to share development costs and achieve similar economies of scale. Cooperative efforts in which vendors work with their channel partners are a reversal of earlier EDI models, in which large companies set all the ground rules, and individual suppliers had to accept those rules or take their business elsewhere.

The Internet broadens connectivity. VAN providers typically run networks that link members only. A company with several trading partners on different VANs might need different software and network connections for each. Today, organizations are piggybacking EDI onto their existing networks rather than having to use VANs.

The Internet reduces the cost of using EDI. The cost of access to the Web is a reasonable monthly fee, and there is no cost per transaction.

19.2.3   EDI Boutiques

An EDI boutique is a company that offers EDI translation software and a network via its Web site—in effect, an EDI service bureau of old.

EDI boutiques are ideal for organizations that expect to send only 40 to 50 EDI documents a month. Organizations with high volumes should look at implementing Net-based EDI on their own. Some organizations may continue to use their VANs for high-volume, guaranteed connections. The Internet provides an alternative, as well as a link to smaller suppliers.

Many of these Net-based enterprises also offer other services as well. For example, if an organization does not typically maintain an EDI capability but would like to track government contracts available via EDI, the organization can subscribe to the services of an EDI boutique such as Loren Data. The software reviews government RFPs for matches with the organization's criteria. When a match is found, the software converts the EDI document into an HTML (Hypertext Markup Language) file and notifies the company either from the Web page or via E-mail.

19.3   Translation Standards

The United States and Canada use the ANSI (American National Standards Institute) X.12 protocol. Europe uses a standard called EDIFACT. Organizations currently using EDI internationally need to support both standards.

X.12 was designed to support cross-industry exchange of business transactions. The standard specifies the vocabulary and format for electronic business transactions. Although X.12 is considered a standard, there are many variations of X.12—each variation contains "enhancements." As a result, organizations end up running multiple versions of X.12 to enable them to communicate with all their trading partners.

EDIFACT stands for Electronic Data Interchange for Administration Commerce and Transport. This ISO standard was proposed as the worldwide standard for EDI, superseding both the X.12 and Tradacoms standards.

19.4 Security

Security has been one of the Internet's weaknesses and therefore a major obstacle to the widespread adoption of Net-based EDI systems. Transactions carry the risk of financial fraud or theft—transmission must be secure.

To remedy this situation, EDI systems vendors, as well as many E-mail, World Wide Web, and cryptography software vendors, are working on developing encryption and protocols to enable secure Internet commerce. Great strides have been made, but there are still obstacles to overcome—not the least of which is the U.S. government. See Sec. 20.3 for more information on that obstacle.

The original Internet communication programs transmitted data between client stations and file servers in unencrypted ASCII text or plain text. Anyone who happened to be sitting along a message route—such as the operators at a relay station or a hacker—could intercept it and read its contents. An interception could prove disastrous if the message contained a credit-card number or instructions meant for a business's bank.

A company can easily create an extranet by using a commerce server that supports Secure Socket Layer (SSL) and placing sensitive pages or perhaps the whole site on that server. This is a good solution if only small portions of the Web activities need to be secured and if the security needs to extend only from browser to Web server. However, if the goal of the extranet is to give trading partners access to an existing application, a company should consider investing in a virtual private network or tunneling that encrypts all traffic from end to end.

Tunneling works at the IP level, so all traffic is encrypted, not just Web traffic. Many organizations use browser-based encryption schemes, such as Netscape's Secure Commerce Server, but these solutions are not nearly as comprehensive as tunneling software.

Many organizations are using a combination of hardware-based firewalls and tunneling software to secure important intranet communications.

Most EDI software vendors offer encryption of Internet transmissions and impose digital authentication schemes. Their offerings ensure that data received from an entity has arrived unaltered and confirm the identity of sender and recipient. For example, Premenos's Templar product relies on VeriSign's technology to generate a digital signature for each outbound message, verifying the identity of the sender and automatically detecting any alteration of the message upon receipt.

19.5 Internet EDI Security Options

Software developers and the standards-making bodies are also doing their part to simplify Internet-based EDI transactions. Software interoperability is key to the success of Internet-based EDI. Off-the-shelf, standards-based software and Internet-based EDI services will make EDI more available, not to mention cheaper and faster than VAN-based EDI. The end result is that the number of EDI users will increase exponentially.

The IETF's EDI-Internet Integration (EDIINT) working group is formulating standards that will let EDI transmissions travel over E-mail, World Wide Web, and ftp (File Transfer Protocol) connections.

Two drafts expected to standardize Internet-based EDI security on the IETF's S/MIME and Pretty Good Privacy protocols were submitted to the IETF in the summer of 1997. A request for comment (RFC) is expected by the middle of 1998. These can be used to enclose EDI messages for secure, reliable delivery over the Internet. These two standards are discussed in Sec. 11.4.

19.5.1   CommerceNet

CommerceNet (www.commerce.net), an association of software developers, banks, telephone companies, Internet service providers, and other companies involved in E-commerce, is sponsoring interoperability tests between Internet-based EDI software from companies such as AT&T, Digital Equipment Corp., Harbinger Corp., Premenos, Actra Business Systems L.L.C., and CyberPath Inc. Once this project is complete, businesses will be able to exchange documents even if they have standardized on different Internet-based EDI software.

19.5.2   Crossware

Netscape's vision for the Networked Enterprise is Crossware, a new category of on-demand applications that run across networks, operating systems, and platforms, and are based on open standards.

Netscape supports Enterprise JavaBeans as an important addition to the enterprise network computing framework. In contrast to operating system—specific component models, it enables customers to deploy intranet and extranet solutions that leverage existing investments in corporate IT systems, applications, and data.

According to Netscape, Crossware describes applications that are available on demand and that run across networks and operating systems. The set covers directory, security, and software distribution standards and provides a common blueprint for companies to use to create "crossware" applications for linking with customers, partners, suppliers, and distributors over the common Internet infrastructure. Netscape will use Common Object Request Broker Architecture (CORBA) and Internet InterORB Protocol (IIOP) as the standards for these Internet applications.

Netscape's next-generation client and server software, code-named Mercury and Apollo, will support Crossware in 1998. Crossware products will be based entirely on Java, JavaScript, and HTML. Federal agencies are the prime targets of Crossware because it will provide a universal interface for a range of platforms and operating systems, and eliminates the need to convert or port code. See Sec. 8.5.2 for more information on Mercury and Apollo.

The core set of standards supporting Crossware applications includes the Lightweight Directory Access Protocol (LDAP), which is an open standard format for data storage, and S/MIME, which is an open standard for sending encrypted E-mail messages. An initial draft of a white paper is available at home.netscape.com/comprod/at_work/white_paper/extranetstds.html.

The Crossware initiative has been endorsed by Netscape and about 40 other vendors.

19.5.3   Open Buying on the Internet Protocol

The Open Buying on the Internet (OBI) protocol, spearheaded by American Express Co. and startup SupplyWorks Inc., will create an architecture for real-time, mass-volume transactions between businesses and their suppliers for everything from paper clips to PCs. Its creators hope that OBI can lend credibility to EDI transactions in the same fashion that Secure Electronic Transactions (SET) has done for consumer credit-card purchases on the Internet.

OBI combines technologies from both the Internet and EDI, enabling a purchasing agent to link a company's internal intranet directly to supplier Web sites. OBI already is backed by a consortium of companies that includes all major commerce software developers and business supply vendors: Microsoft, Netscape, Oracle, Open Market Inc., BASF, Office Depot Inc., National Semiconductor Corp., United Technologies Corp., and VWR Scientific Products.

By organizing disparate electronic commerce systems around a common standard, OBI could drive down the cost of making business purchases. The OBI specification is directed at business-to-business commerce, incorporating EDI and trading partner relationships over the Internet and extranets. The OBI standard includes purchase order management and links to payment institutions.

One OBI implementation, for example, would enable a supplier to authorize an agent with an X.509 digital certificate and publish a catalog customized for that agent's business. The supplier would package a purchase request into an OBI purchase order, which is based on an EDI document standard called X12-850, and send it back to the purchasing agent.

From there, a company can enter the order into a workflow system for approval or link to external payment providers such as American Express's Corporate Purchasing Card system.

19.6   Net-based EDI Software

With Net-based software, business document exchange should be simpler and less expensive than with traditional EDI PC software. Trading partners don't have to learn about X.12 or EDIFACT. EDI processes and communications testing should be minimal.

Net EDI software falls into one of two categories. Web EDI uses secure Web sites to translate documents between the HTML format used by the Web browsers and the standard EDI format. Transmission can be over the Internet or over a private network provided by the software vendor. The other category uses existing standards for Net E-mail, encryption, and digital signatures.

EDI providers such as GEIS, Harbinger EDI Services, and Sterling Commerce offer products that allow organizations to implement Net-based EDI on their own.

19.6.1   EC Exchange

In mid-1997 the EC Company (www.eccompany.com) launched a virtual private network, the EC Network, over which it transports EDI transactions. EC Exchange software manages electronic payments and time stamps. It also archives documents and provides the Internet connections linking trading partners to UUNET Technology Inc.'s backbone. EC Start enables companies to receive business transactions and payments from EC Exchange users over the EC Network.

The service also helps translate from one document type to another, simplifying one of the most confusing aspects of EDI. EC Company has developed a metalanguage of standard data elements, such as purchase-order headers and purchase-order decode files.

19.6.2   Netscape ECXpert

Actra Business Systems, Inc., a joint venture of Netscape Communications Corp. and General Electric Information Services (GEIS), offers a suite of servers as part of its CrossCommerce initiative for corporations looking to sell to consumers and business partners over the Internet. The suite includes ECXpert, OrderXpert Seller, OrderXpert Buyer, MerchantXpert, and PublishingXpert. Netscape bought out GEIS's interest in the venture in late 1997, and now has full ownership of Actra. All former Actra products are now known as Netscape products.

Netscape ECXpert provides companies with the ability to interface and work with EDI systems over the Internet. ECXpert provides EDI mapping translation to back-end systems, document tracking, and trading partnership management. Management of ECXpert is handled by a Java-based application that enables users to establish cross-company relationships, integrated workflow, and access control.

Actra builds much of its technology around CORBA and IIOP, which will enable the servers to share resources, features, and capabilities.

Actra leveraged Netscape's technology from its Commerce Server and Publishing System and GEIS's experience with EDI and virtual private networks. Actra took over the stewardship of Netscape's Merchant System and Publishing System when the company was formed and released an update to both products in late 1997. While the Xpert line of servers will be aimed at the business-to-business market, Actra revamped Merchant System and Publishing System to connect businesses to end users by making the systems able to handle hundreds of thousands of hits per day.

Netscape Publishing Xpert offers document management, replication, push delivery, and one-button publishing. Netscape Merchant Xpert offers greater flexibility for storefront presentation, customizable back-end services, order-status monitoring, customer profiles, and dynamic catalog presentation. Merchant Xpert APIs are being defined that will enable corporations to easily connect the server to order-entry systems and to inventory management and customer service applications.

19.6.3 GE TradeWeb

GEIS is one of the largest VAN providers. It is offering GE TradeWeb, which is a subscription-based service similar to Harbinger Express. TradeWeb is targeted toward small businesses.

GE TradeWeb allows Internet users to access a secure Web server to retrieve EDI documents that have been translated into HTML, and to access forms that can be translated into standard EDI and forwarded to trading partners on GEIS's VAN. GE TradeWeb allows large companies using GEIS's VAN to reach smaller trading partners that haven't used EDI in the past.

GE TradeWeb translates forms into X.12 EDI formats and sends them off to trading partners. Transaction security is provided through SSL encryption.

GEIS maintains subscribers' EDI mailboxes on the GE TradeWeb server, as well as a directory of companies that can be reached using the service and the forms needed for specific business transactions. Subscribers pay a small initiation fee. For a monthly fee or a flat annual fee, users can send up to 30 documents a month and receive an unlimited number. They're also entitled to technical support by telephone or E-mail. Infrequent users can pay a fee per document sent.

19.6.4   Harbinger Express

Harbinger Corp. (www.harbinger.com) offers Harbinger Express—an example of Web EDI software—which allows companies to use browsers to exchange documents, which are then automatically translated from HTML to EDI format. Harbinger sets up the translation mechanism on a case-by-case basis between trading partners. Once the partners have been "certified" for trading, they can use Harbinger Express to translate documents such as invoices and purchase orders into a form that can be read through standard Web browsers.

19.6.5   Templar WebDox

Templar WebDox from Premenos Corp. (www.premenos.com) uses E-mail and the Web for internal and external document exchange, using the SSL protocol to allow an organization to transmit EDI forms over the Internet. Businesses at each end can view documents via a Web browser. WebDox can also be configured to work in conjunction with on-line catalogs. When a customer orders a product from a company's on-line catalog, WebDox generates the appropriate forms and transmits them to the company. WebDox consists of two major elements:

- WebDox Central, which resides on the server of a large organization
- WebDox Remote, which resides on the PCs of smaller trading partners

19.7   Success Stories

Most organizations that have turned to the Net for their EDI applications have been able to cut costs quickly while improving efficiency at the same time. The stories that follow are but a few of the successes. They were chosen to show that success is not limited by industry or size.

Wal-Mart   Probably the most familiar success story for Internet-based EDI is Wal-Mart—the "Every day low prices" store. Wal-Mart is focusing on achieving productivity gains through technology and other innovations. The company very quickly embraced the idea of an Internet-based EDI with its RetailLink system and insists that its business partners—including manufacturing partners—adopt the same standard.

But Wal-Mart has gone beyond the original idea of EDI—electronically exchanging data that was on paper. Wal-Mart recognized the major opportunities for applying E-commerce to supply-chain activities such as planning, purchasing, and logistics. For example, by sharing up-to-the-minute demand forecasts with suppliers via extranets, Wal-Mart has said it expects to cut inventories and improve in-stock performance.

Warner-Lambert, the maker of Listerine and other consumer goods, recently wrapped up a collaborative planning and forecasting "paper" pilot project with Wal-Mart. The closely watched pilot, which relied on no new technology but simply reworked planning and forecasting processes to include collaboration, showed the potential to increase the availability of merchandise in stores from 87 percent to 98 percent while reducing inventories and reducing the time it takes to get products onto store shelves from 12 weeks to 6 weeks. The reason: Wal-Mart had a better handle on store-level activity and the impact of competition from other retailers, while Warner-Lambert's forecast had more complete information on the impact of promotions and new products. Combining them allowed both companies to predict demand more accurately.

Chrysler's SCORE   The system Chrysler developed, known as SCORE, for Supplier Cost Reduction Effort, almost overnight helped Chrysler slash billions from its manufacturing and business operations. The second phase of this simple electronic-commerce application, released in mid-1997 over an industrywide intranet, is expected to cut costs by $2 billion annually by 2000.

Since more than 70 percent of the parts found in Chrysler cars are purchased from outside suppliers, SCORE was originally implemented as a way to improve relations between the then-struggling Chrysler and the vendors that provide it with everything from subassemblies to brooms. The program soon incorporated incentives to motivate suppliers to suggest products that could help the auto manufacturer reduce its costs. Not only did the supplier who suggested a cost-saving approach earn Chrysler's business, but it was also rewarded with a portion of the savings achieved.

SCORE runs using a single Notes 3.0 application, one Notes database running on a SunSPARC 2000 workstation and roughly 760 Notes clients, 160 of which are on the desktops of suppliers. At the moment, the system doesn't even include E-mail.

The key was that Chrysler designed the business process and then applied the technology so as to collect all the information in one spot and allow all the appropriate parties to look at it.

To make it easier and more reliable for suppliers to access SCORE, Chrysler will move the system onto the Automotive Network eXchange, a private industry network developed by the Automotive Industry Action Group. The network is a collaborative effort on the part of the Big Three auto makers and several hundred of their suppliers. ANX will function like a giant intranet, providing multiple pathways, high-level security, and network redundancy between Chrysler and its suppliers.

SCORE 2, which is expected to be rolled out during the summer of 1998, will give all Chrysler employees access to Notes 4.0 and Notes Mail, which means that the entire process can be moved on-line. Notes security features, including encryption and author-level access rights, already ensure that suppliers can view only their own documents.

Fruit of the Loom   Fruit of the Loom extended its network to distributors and their decorator customers, such as T-shirt embroiderers and screen printers. Distributors and decorators can now use the Internet to access on-line catalog and order-processing capabilities. The distributor becomes the hub access point to Fruit of the Loom, providing critical marketing and sales activities. Decorators now have visibility into what inventory is available, what substitutions are possible, and what customer-specific specials and promotions are running, and can view distribution and transportation information.

Daisytek   Daisytek International is a leading wholesale distributor of computer and office automation supplies and accessories. The company distributes more than 8600 products from more than 145 manufacturers to its resellers throughout North America.

In early 1996, Daisytek rolled out an on-line ordering system dubbed SOLO (System for On-Line Ordering) designed to increase the efficiency of the product selection, ordering, and fulfillment processes of its partners. SOLO enables resellers to answer customer inquiries immediately while dramatically reducing the number of calls to Daisytek to confirm pricing and product availability.

Resellers simply click on the SOLO catalog information, which is supplied on CD-ROM, build their purchase list, and communicate with the SOLO database via the Internet. The database is automatically updated when any changes in pricing or availability occur. Resellers can close orders with a single call and can integrate the order-entry transactions with their own back-office systems.

Owens Corning   Owens Corning Inc. began testing Web EDI in mid-1996 and introduced it to 100 trading partners in mid-1997. Developed with Sterling Commerce Inc., its VAN provider, the system lets distributors view Owens Corning's products and prices, and place orders on a secure Web site with a browser. These orders are encrypted for Internet transmission and then converted to EDI format by Sterling, which then forwards the orders to Owens Corning.

D&T Distributing   EDI is used by only a small percentage of D&T Distributing's customer base, but it accounts for nearly a quarter of D&T's business. D&T's entire catalog is available on EDI. To D&T, that means eliminating paperwork and not having to have someone rekey the purchase orders. When a retailer places a purchase order, it gets printed at D&T; automatically entered into D&T's back-end systems; and then shipped, tracked, and confirmed.

Lee Printware   Lee Printware Inc. uses EDI for links to its larger retailers and sees no reason to move those links to the Web yet. Lee was using EDI to link to its distributors but found it too expensive and time-consuming. Web to the rescue. Distributors now use their browsers to link to Lee Printware's TieNet and check product availability, place orders, and check order status. All customer account and shipment data is on a mainframe inside the corporate firewall. All transactions are encrypted with SSL.

Mobil Corp.   Mobil had two different EDI systems, one DOS-based and one Windows-based. VAN charges for the hard-wired networks were topping $100,000 a year. Every time Mobil changed a business rule, new software had to be sent to all the dealers and installed on their desktops. Inventory information was updated only once a week. Dealers began to resort to phone calls and faxes.

The new Web-based EDI does not require a hard-wired network, so VAN charges disappeared instantly. When business rules are changed, they are placed on the server and are available immediately to all trading partners.

The business rules are embedded in the system's Java applets. The system immediately alerts a distributor if the entered order is incorrect. Distributors get up-to-date inventory information. In addition, they can review order status and order history. They can enter partial orders and place them on hold.

The Net application has helped the company realize its goal of a "perfect order," one that comes in and gets processed correctly—and quickly—without a great deal of intervention.

19.8 The Future of EDI on the Net

Some analysts predict that Net-based E-commerce products will offer much more flexibility and be more attractive than traditional EDI and that EDI will die a slow death. Others say that they are just waiting for the security to be beefed up before they enter this new arena. As with any business-critical application that is based on the Internet, organizations have to consider the major link in the architecture—the Internet itself. It is crowded, slow at times, and underfunded. It is owned by no one.

But even with that uncertainty, organizations are moving their EDI to the Internet. The Internet technologies allow EDI to be layered over existing applications. The standards base enables interoperability.

With traditional EDI, adding a trading partner means delivering a turnkey system to that partner's site. Those systems cost in the thousands of dollars. With Net-based EDI, once the application is built and sitting on the server, adding a new trading partner means adding some user names and passwords. After that, all it takes is the partner's browser.

Most new suppliers will use the Net rather than a VAN. The VAN business model will shift from a transaction orientation to a service emphasis, with guaranteed bandwidth and end-to-end integration.

Distributors will become more relationship-oriented rather than transaction-oriented. The focus will change to helping the customer reduce the total cost of doing business.

The Web may just be the push EDI needs for wide-scale acceptance. Only time will tell.

20

E-Commerce: Where Is It Headed?

Electronic commerce has the potential to affect every industry and every global market. For many analysts, the question is "when" rather than "if." Organizations need to rethink their approach to business—to think out of the box, to use a reengineering term.

Peer-to-peer electronic commerce currently has the most momentum. This removal of middlemen—disintermediation—has flattened the economic value chain. People have easy access to vast amounts of information. The resulting reduced barriers to entry drive free market dynamics and lower prices and margins.

With the connectivity of the Internet, markets can become truly global. Using directories and certificate authorities, users can quickly and easily find and investigate potential trading partners.

New electronic payment systems and services support global peer-to-peer payment with minimal overhead. As businesses easily disperse or outsource operations or relocate, governments have difficulty regulating borders, taxes, and currencies.

The face of EDI is changing. Originally built by and for the exclusive use of large vendors, EDI networks will now be offered as commercial services. Many companies will outsource to large systems integrators rather than build their own private network. Ultimately information—rather than the actual transactions themselves, as is the case today—will be the major flow between trading partners.

New businesses designed around the Net will have leaner cost structures. Inventories will be reduced substantially by just-in-time and build-to-order manufacturing. Organizations will have electronic links to both upstream manufacturers and downstream third parties that manage delivery logistics. Goods will delivered inexpensively to a local pickup outlet or shipped directly for a fee.

E-commerce becomes what an organization chooses to make of it. CompUSA has a Web site that allows consumers to easily compare different products side by side. Wal-Mart recently put tens of thousands of computer-related SKUs available for sale on-line. Ingram Micro is providing the fulfillment so far, but other distributors are also talking to Wal-Mart.

Successful relationships have retailers, distributors, and vendors working together. As long as those involved continue to work together to make the partnership work, each benefits. Talk about a new way to do business!

20.1   Efficient Consumer Response

Retailers are under pressure from shrinking profit margins, intensified competition, more restrictive consumer legislation, and more demanding and sophisticated customers with reduced brand loyalty. A new strategy called Efficient Consumer Response (ECR), with EDI at its foundation, offers hope in cutting operating costs in the supply chain.

The guiding premise of ECR is that retailers and their suppliers must work together more closely to bring better value to the consumer. This goal can be achieved by focusing on the efficiency of the total supply chain rather than on individual components. For this to happen, all members of the supply chain must be willing to work together. EDI can be used to effectively automate the flow of information among the supply chain partners.

The traditional supply scenario was that manufacturers set their targets on the basis of what they thought demand from their customers would be.

Based on these assumptions, they ordered raw materials and began production to meet the forecasted demand. If the forecasts were too high, they ended up with surplus product. If the forecasts were too low, they ended up trying to produce more to eliminate shortfalls caused by the unusually high demand. In other words, how much to produce was always a guessing game.

Today the retailers are the ones in charge. They are demanding that their suppliers track their needs and automatically supply goods according to preset thresholds. They are expecting suppliers to warehouse the inventory so that they don't have to; they are expecting shipments to their individual stores, not to their own warehouse, from which they would have to ship the products to their individual stores. In addition, customers are expecting suppliers to provide more services, such as prepacking shipping pallets with items from several product lines.

The key to ECR is effective use of information technology. For example, point-of-sale terminals capture data during the day. At the close of business, all stores send the data directly to suppliers. Headquarters' machines send data from any on-line orders that occurred that day. The supplier's EDI software takes this information and determines what can be sent (and to where) and what needs to be flagged as back-ordered, sends back confirmations, generates purchase orders for raw materials needed to produce the back-ordered items, and sends those purchase orders to its own suppliers.

One benefit of ECR is higher customer satisfaction and responsiveness. The biggest payoff is in reduced waste and excess inventories, which in turn reduces costs. Suppliers know exactly what is needed to replenish stocks. The store carries fewer weeks of inventory. Suppliers can better forecast their needs and also carry fewer weeks of inventory.

20.2   Smart Cards

"Smart-card" applications are being developed that hold not only a user's profile and network-access privileges, but also financial and health-care information, and can also be used for shopping over the Internet. A smart card is a credit-card-like device with an embedded microchip. It slips into a smart-card reader that is standalone or installed in a keyboard. Information to support applications such as network security and electronic shopping is stored on the chip.

Smart cards may replace wallets—the term applied to software that equips browser users with digital certificates and payment systems.

Smart-card development has been hampered by the lack of standards. The ISO 7816 smart-card protocols specify standards for card sizes, placement of contacts for reading card contents, and interactions between cards and card readers. The EMV specification developed by Europay, MasterCard, and Visa has become a de facto standard for electronic payment systems. The EMV specification ensures that all smart cards operate across all card terminals and related devices regardless of location, financial institution, or manufacturer.

Analysts feel that the acceptance of crypto smart cards has been hampered by a lack of agreement on standards. Regardless of how this turns out, the early adopters of this technology will be the financial institutions already doing business on the Web.

See Sec. 11.3 for more information on smart cards.

20.2.1   Smart-Card Examples

IBM has a smart card based on the Secure Electronic Transactions (SET) protocol operating in a variety of applications, including cash purses (areas on the smart card's silicon to store digital cash) and airline and hotel ticketing. Using the ticketing application currently being tested by American Express Co., American Airlines Inc., and major hotel chains, users download airline and hotel reservations over the Internet, and these are stored, along with cash, on the card.

VeriFone offers Personal ATM for digital cash. The user slips the card into the device, makes a connection with the bank, and keys in the amount to be downloaded. The card is then used to make purchases on any payment terminal that will accept it (like ATM machines) or over the Internet with a reader connected to a computer.

20.3   Stumbling Blocks

Some believe that E-commerce is going to revolutionize the way organizations do business and that this will happen soon. Others recognize that there are great benefits from this new mindset but are pessimistic about how quickly it will take over, citing a number of areas that aren't mature enough yet.

20.3.1   Image

For companies implementing electronic storefronts, there are some major differences between setting up a physical store and setting up an electronic store. When customers walk into a store, they can see the products, touch the products, walk around the products, ask questions about the products, etc. The salespeople don't give a textbook answer to each question—they are able to tailor their answers based on their perception of the customer. Also based on that perception, they can give additional information that they think will help close the sale.

In an electronic store, even if the products are viewed with 3D software, customers can't feel the products. Any questions they have must be answered either from a set of canned responses, or using a search engine on the site, or via E-mail to the company, which means that the customer waits for a response—usually for more like days than minutes. There is no way the computer application can size up the customer and figure out what it will take to make the sale.

Therefore, companies need to realize that what they actually selling is information about their products, not the products themselves. It is the amount of information, the completeness of the information, the ease of retrieving information, and so on that will ultimately help make the sale.

20.3.2   Technology

The technology is there, no question. It's no coincidence that the companies that stand out for their E-commerce applications are either, in the case of EDI, already into EDI technology, or, in the case of electronic stores, already knowledgeable about technology because that is their industry. Companies that have started up as electronic stores carry no baggage, so developing the infrastructure is easy—they are starting from scratch rather than trying to retrofit legacy (or client/server) applications into this new technology.

20.3.3   Bandwidth

For electronic commerce to be successful, messages need to travel around the Internet reliably and in a timely manner. When a shopper places an order, it should arrive almost instantaneously. When an order is placed with a supplier, it should arrive almost instantaneously. In addition, the message sent should be the message received. One of the largest hurdles when using the Internet today is clogged bandwidth. For E-commerce to be successful, the bandwidth is going to have to support more messages.

20.3.4   Security

Security continues to be an issue. Acceptance of SET should calm these fears. Credit-card numbers will be at least as safe as they are when they are given over the phone.

IT organizations must understand and accept public-key cryptography. Information-intensive businesses such as banks need to develop secure content management architectures that take advantage of emerging Web standards. Most organizations need to upgrade their communications and storage infrastructure to accommodate high-bandwidth applications. Marketing-oriented E-commerce applications need to take advantage of new media forms and more wireless connectivity.

Encryption does not stop interlopers from capturing messages. It makes it (hopefully) difficult for them to decipher these messages. Encryption is based on mathematical equations, and with today's computing power, messages should be encrypted so well that only the intended recipient with the key should be able to decipher them.

However, encryption is a cumbersome task and takes time. In addition, the U.S. government strongly discourages the private use of high-quality cryptographic software—even going so far as to classify encryption programs as "controlled munitions."

SSL was cracked by a European hacker using an array of supercomputers, but the hacker cracked the international version, which uses only a 40-bit encryption algorithm—the U.S. government forbids the export of any cryptography software that is any stronger.

Microsoft's Private Communication Technology (PCT) protocol, which is included in Microsoft's Internet Information Server 3.0, is an enhanced SSL. PCT uses a second key specifically for authentication, but this key is not restricted by the government and uses a superior random number generator.

Momentum System's F-Mail allows the secure automated transfer of EDI documents using the standard Internet ftp (File Transfer Protocol) program. F-Mail operates much like an E-mail system, but encrypts files with ViaCrypt, the commercial version of PGP. F-Mail can also confirm the delivery of messages through the generation of a return receipt. In addition, it can be used with a security callback option when connected to an Internet service provider (ISP).

While discouraging encryption for the private sector, government agencies use a mandatory security solution known as Fortezza. Fortezza encrypts transactions using a minimum 56-bit encryption key based on Data Encryption Standard (DES). Each government Internet user is assigned a unique identification string which is stored on a credit-card-sized micro-processor called a token. These tokens function as personal Internet keys and are carried by each federal employee. All Internet workstations and Web servers are equipped with card readers into which the authentication tokens are inserted. To use them, users must enter a password.

Joint Electronic Payment Initiative Another security protocol on the horizon is the Joint Electronic Payment Initiative (JEPI), developed by IBM, Microsoft, Visa, and MasterCard as an alternative to SSL. JEPI works in conjunction with SET.

JEPI has two parts. The first is an extension layer called PEP (Protocol Extension Protocol) that sits on top of a Web server's basic Hypertext Transfer Protocol (HTTP). The second part, UPP (Universal Payment Preamble), is a proposed negotiation standard for different payment methods such as SET-enabled software and smart cards.

20.3.5 Size of the Market

Statistics abound regarding how many people have access to the Internet, making them potential customers for an electronic storefront. But these numbers include people who can access from their local library and students who can access from their school.

A bigger question is how many of the people who have access to the Internet would buy over the Internet. For some, it is a question of security. For others, the Internet is still just for information—these people may go shopping on-line but actually purchase off-line.

Ask 10 people you know who regularly surf the Internet. How many of them have actually bought something—anything—over the Internet?

Early adopters are speculating that the number of people willing to buy over the Internet will grow exponentially. They want to be in place and have name (Web) recognition when that happens. Look at Amazon.com. Its initial public offering was incredible, and it had yet to make any profits.

20.3.6 Digital Payments

The cost of processing digital money is still too high. Acceptance of Digital's Millicent and CyberCash's CyberCoin technologies may generate a new on-line industry that allows customers to pay for exactly what they use. An information lookup might be 5 cents, an applet download 10 cents, to listen to a song on a musical CD 25 cents, to play a game once a dollar. As the cost of digital payments goes down, the possibilities go up. On-line magazines with ads might be free, but advertisement-free versions might cost 20 cents. Site visitors might get paid for providing information about themselves. Software publishers could charge small amounts for features or upgrades.

20.3.7 New Channel, New Software

After all the push toward client/server architectures, E-commerce assumes thin clients and a host-centric environment. Applications have to be written (or rewritten) for that approach, and automated routines need to be created to monitor the efficiency of on-line advertising.

20.3.8 Disintermediation

The Internet can be used to remove the middleman—and take out layers of markup and cost—and totally alter distribution channels.

Distributors are especially at risk of being cut out. To shore up their worth, distributors need to provide postsale services, add value by supplying information such as warranty terms and inventory availability to their customers, and add value through outsourcing manufacturing and on-site logistics from companies producing the goods. Distributors need to put themselves in their customers' shoes and reorganize from the outside in.

Larger chains will use distributors for entirely different reasons from smaller retailers. Smaller retailers will play a larger role in determining how distributors use the Internet as a marketing tool. Retailers have turned to distributors to deliver product as needed rather than storing inventory in their own warehouses. Distributors that can't provide the products lose the customer. As a result, the larger a retailer is, the more likely it is to have a close relationship with one or two major distributors with which it has forecast its requirements.

20.3.9 Brand Identity

Marketing 101 taught us that brand identity and brand loyalty are important factors in maintaining one's market share. For some E-commerce pioneers, this continues to be true. Dell Computer's success could be attributed in part to brand identity.

However, the flip side of brand identity and loyalty is price sensitivity. And on the Internet, price comparison is just a click away.

20.3.10 Changing the Way Organizations Do Business

"Build it and they will come" is not true of E-commerce. EDI ties companies together even more closely than before, which changes the nature of the business relationship. Companies that tie their information systems together share accurate information in a timely manner, which helps get products to market quickly and inexpensively. Inventory production and distribution applications that work off real data rather than estimates should lead to greater efficiency.

But organizations cannot force business partners to use the new system. Customers and suppliers alike have to see and experience the benefit. And they can afford to be choosy. FedEx originally offered free tracking software that used proprietary software. The company found out that even free proprietary software was a no-go and finally offered an open tracking system on the Internet that has become widely used.

An organization can't just take current processing systems and simplify them for the Net. Order takers have access to data on all a company's customers, but customers entering the system through E-commerce shouldn't have access to one another's data.

20.4 Outstanding E-Commerce Issues

The ultimate success of electronic commerce may actually depend on the resolution of several critical international legal issues. Questions abound regarding the location and rate of tax levies for goods sold via the Internet. In addition, individual countries often have conflicting laws regarding intellectual property and unauthorized use of Internet content.

20.4.1 Taxes

The Framework for Global Electronic Commerce, released by a Clinton administration policy committee, is a statement on electronic commerce that suggests that the federal government will treat the Internet as a tax-free zone. The document covers a range of financial and legal issues. The document urges international cooperation on copyright laws and freedom of the press issues. Encryption technology, which the government has vigorously sought to regulate, claiming national security risks, is still restricted. Electronic banking standards and other technological issues are left to the private sector, citing the industry's rapid rate of technological advancement. The policy says nothing about state taxation.

While the federal government may be promising a hands-off approach to electronic commerce, state governments, which receive over 30 percent of their revenue from sales taxes, are not. Interstate commerce is generally not taxed unless the seller has a nexus, or a significant sales or physical presence, in the buyer's state. This has allowed many large mail-order companies to deal with state tax in only a few states.

However, with transactions conducted over the Internet, several states could be involved. In fact, the definition of *nexus* becomes the critical issue. Is an ISP considered a nexus, and thus a reason for taxation if the ISP and the customer happen to be in the same state? If the ISP can be considered a taxable sales presence, then the telephone company, as owner of the leased lines, could also be considered a reason for state taxation. The argument is that since the physical server receiving the on-line purchase actually contains the transaction, then the state in which the server is located should be paid state taxes.

States are also looking at other Internet-related ways to generate revenue, such as business license taxes, Internet-access fees, franchise fees, gross-receipts taxes, excise taxes, privilege taxes, and utility taxes. These fees would be paid in every state in which a company installs a Web server. So a bookseller doing business on the AOL network would have to collect taxes wherever AOL has a modem or server.

Connecticut, New York, and Texas already have broad-based tax schemes for computer services and consider Internet-related transactions as computer services. In Connecticut, a "point of presence" such as an access modem in the state is enough to create a nexus. In Texas, a Web page based on a server located in the state creates a nexus. The rationale is that if software is downloaded from a server in Texas, it's the same as coming to Texas to pick up the software.

California is much more friendly. California lawmakers decided that having a Web page in the state does not create a nexus. By creating an Internet tax-free zone, the state is hoping to attract economic development.

Internet Tax Freedom Act   The Internet Tax Freedom Act, which proposes a moratorium on taxes on electronic commerce and the repeal of taxes already placed on Internet service providers, has been introduced in Congress. Updates on this law can be found at www.house.gov/cox/nettax/taxfreedom. Basically, the bill states that no tax will be imposed by any state, county, or municipal taxing authority on Internet or interactive computer service activity. One sticky distinction is that between goods that are ordered over the Net and goods that are ordered and delivered over the Net, such as downloadable software or data. However, given the large amount of revenue at stake, it is likely that the states will take the issue to court—although some question whether the taxes collected on Net-delivered goods will be worth the effort.

The Multistate Tax Commission, which represents 41 states and the District of Columbia on tax issues, is exploring whether on-line sales tax collections can be shifted to credit-card companies or electronic cash companies. They are also looking to see if tax compliance could be embedded in electronic commerce software.

An interesting side note is that there seems to be little or no effort to impose value-added taxes on the back-end, business-to-business portion of E-commerce. So EDI seems to be out of the hot seat—for now.

20.4.2 Privacy

Much is said about securing credit-card transactions. What about other information that goes with the transaction—addresses, social security numbers, buying habits, etc.? How do users know that they can trust the merchant not to use or sell that information?

20.4.3 Authentication

Sounds good on paper, but how does a merchant know that the user didn't get the credit-card number off a discarded charge slip?

20.4.4 Liability Issues

Who is liable for failed encryption or authentication? What if material on a Web site to which your site links is illegal—is your site liable? If a message is delayed and a company doesn't get its supplies on time, is the Internet service provider liable?

20.4.5 Contracts

To what extent do current commercial and contract laws apply to electronic-based contracts? Is an electronic-based contract with digital signatures binding? Electronic contracts "signed" digitally in one state may not be enforceable in others. Utah does not recognize digital signatures; New Hampshire does.

20.4.6   Consumer Rights

What methods will be used for transaction settlement on the Internet? If the merchant is responsible for the effects of a virus downloaded with software, how does the consumer collect? Is it the merchant's responsibility, the owner of the receiving machine's responsibility (to check for viruses), or the communications vendor's responsibility if the virus was tagged on during transmission?

   If a user is treated unfairly by an on-line merchant, how does the user track the merchant down? How do users know whom they are dealing with and whether merchants will do what they say they will?

20.4.7   Burden of Proof

Our court system relies on proof. With transactions happening instantaneously and with no paper trail, what should organizations do? Should they archive the electronic version of the transaction? Will that be accepted in a court of law?

20.4.8   Legal Issues

Organizations that offer encryption software are not allowed to sell it internationally or seek international partnerships. Pretty Good Privacy Inc. has found a way to get around that restriction. PGP paired up with Schlumberger to market the Cryptoflex smart card. Schlumberger is based in France but maintains a U.S. office. PGP's technology is being manufactured into Schlumberger's products in France—hence the legality of this collaboration (for now, at any rate).

   Industries such as publishing, entertainment, and software are at special risk with E-commerce, as the incremental cost of making a perfect digital copy drops to zero. New technologies for electronic copyright protection and usage management are now making it possible to securely deliver digital goods directly over networks and to get paid through third parties.

   In many professions, information and expertise are coded into the software, such as Intuit's Quicken CD. This trend affects information-centric service industries, including law, travel, insurance, accounting, and real estate. The Net allows providers to extend services globally and allows others to leverage this knowledge. People will turn to the Net for intelligence rather than expensive local individuals (who probably access the same sources!).

Who or what has jurisdiction in cyberspace? A Missouri court found that it had jurisdiction over a California resident, but a New York court found that it had no jurisdiction over a Missouri resident.

20.5   Is E-Commerce Here to Stay?

EDI is being adopted as organizations try to find ways to cut costs and improve customer service. Converting to the Net has many immediate cost benefits. E-commerce technology can be used to reengineer the corporation. As E-commerce features are adopted by small as well as large trading partners, new business processes become possible. By reducing the clerical workload and eliminating unnecessary paper handling, E-commerce can ensure rapid, accurate, and secure exchange of business information; reduce inventory costs; and improve speed of ordering, delivering, and paying for goods and services.

Security becomes a bigger issue for acceptance of EDI. Even with several reliable encryption techniques in use, organizations are still not convinced that transactions are secure. EDI VAN providers with Web-enabled interfaces will continue to flourish until these organizations are convinced. Once the security issues are resolved, more and more companies will begin to use the Net to maintain relationships with their trading partners.

Organizations are not rushing into electronic commerce for retailing, usually citing security issues. But there really are bigger questions: Are customers ready to buy using the Web? Can a company really make money using the Web?

Most businesses now have reserved domain names and have set up Web pages to keep in touch with the masses. It is hard to find a piece of advertising in print or on TV that does not include a Web address. Children as young as 5 know that www.something means the Internet, although they don't necessarily know what the Internet is. And even with all this exposure, only a small percentage of U.S. businesses are actually doing business over the Internet.

Companies that have traditionally offered mail-order and/or catalog sales have gravitated toward the Internet. Customers are already used to buying based on a picture and written description and receiving their merchandise days later, not to mention giving their credit-card number out over the phone. To these customers, Internet buying is just another way to do the same thing—except maybe with the addition of motion.

But are the mainstream retailers taking this seriously? Some are putting up electronic catalogs but don't include all their products. Are they too distracted by the Year 2000 issues?

The push for E-commerce is actually coming from consumers. Yet, cybermalls don't seem to be a huge success. Those who indicate that they would buy over the Net are those that are already comfortable with the technology. Those using the Web are still a small percentage of those with buying power. Key questions remain:

- What are consumers willing to buy over the Net?
- Will more consumers be willing to buy over the Net? When?

GLOSSARY

Active desktop   A version of the desktop for Windows 95 and NT that integrates Microsoft's Internet Explorer into the desktop, allowing users to access their own files and the Web using the same interface.

Active scripting   A language used to control the behavior of server ActiveX controls and/or Java applets; an example is VBScript.

Active Server Page   A file that contains HTML, possibly ActiveX controls, and VBScript or Jscript and generates on the fly a virtual HTML page to be sent to the client; it requires Microsoft Internet Explorer 3.0 or greater or a Navigator plug-in offered by Netscape.

ActiveX   A technology that enables software components to interact with one another in a networked environment regardless of the language in which they were created (Microsoft).

ActiveX controls   Interactive objects in a Web page that provide interactive and user-controllable functions.

ActiveX documents   Documents in a variety of formats automatically linked to a Web site which allows non-HTML documents to be viewed through a Web browser.

API (application programming interface)   A set of programming routines that are used to provide services and link different types of software.

Applet   A small program that performs a specific task.

Application gateways   Network nodes used to route application-specific traffic to a particular application server.

Application program interface   See API.

Application proxies   Act as servers to the application clients and as clients to the application servers; evaluate network packets for valid application-specific data; a type of firewall.

ASP   See Active Server Page.

Asymmetric encryption   Use of public/private-key pairs for data encryption.

ATM (asynchronous transfer mode)   A technology that creates a virtual circuit among two or more points at high speeds; it uses fixed-length packets with header and information fields.

Authentication   That part of ActiveX that allows a client program to dynamically invoke the processing in an ActiveX object; execution of objects (Microsoft).

Automation   Identification of an entity such as a machine on the network.

Backbone   A network that connects other networks.

Bandwidth   The capacity of a communications channel.

Blanket push   Pushing information to users without benefit of filters or user selection.

Byte code   An intermediate language that is executed by a run-time interpreter.

CDF   See Channel Definition Format.

CGI (common gateway interface)   A protocol used by Web-server software to communicate with other applications on the server or on the network.

Channel   A logical grouping or category of information offered by a content provider such as Reuters, PointCast, or CNN.

Channel Definition Format (CDF)   A text-based format that allows push channels to point to Web pages.

COM (Component Object Model)   A component object-oriented architecture that encapsulates commonly used services and functions; encompasses everything previously known as OLE (Object Linking and Embedding) Automation (Microsoft).

Common gateway interface   See CGI.

Common Object Request Broker Architecture   See CORBA.

Component   A reusable software module that is a collection of business objects that handle processing, encapsulate data, and provide necessary user interfaces.

Compound document framework   A framework that facilitates the creation of documents that contain a combination of data structures and may contain pointers to external files.

Container   The application a component is designed to be used within.

Content   Term used for the material (information and data) accessible and displayed by an i-net application.

CORBA (Common Object Request Broker Architecture)   A technology that provides portability and interoperability of objects across heterogeneous systems (Object Management Group).

Cost of ownership   All costs associated with owning a piece of computer hardware; usually used to evaluate client machines.

Cybermall   An electronic shopping mall which links to electronic storefronts.

Data Encryption Standard (DES)   An algorithm for encrypting (coding) data designed by the National Bureau of Standards.

Data mining   Discovering meaningful correlations, patterns, and trends by digging into large amounts of data using artificial intelligence and statistical and mathematical techniques.

Data warehouse   A repository of data, usually from production systems, that is summarized or aggregated and possibly translated; it has an enterprisewide focus.

DCOM   See Distributed COM.

Decryption   Translation of encrypted data back into the original form; the reverse of encryption.

Desktop Management Interface (DMI)   A common format for presenting and retrieving management information for desktop machines.

DHCP (Dynamic Host Configuration Protocol)   A protocol that dynamically configures client address as they are required.

Digital cash   Virtual cash used for payments over the Internet.

Digital certificates   A method of verifying to both parties in a transaction that the holder is the person or organization it claims to be.

Digital signature   An electronic signature that cannot be forged.

Disintermediation   The decline of intermediary companies that today operate between the buyer and the maker of goods—as a result of the Internet, many companies can now do business directly with their customers.

Distributed COM (Distributed Component Object Model)   Used to design applications as a set of components that reside on different machines in a network (Microsoft).

Distributed computing   Systems networked together, rather than decentralized with no communication between them; applications and their components and data are dispersed throughout the enterprise; previously referred to as distributed processing.

Distributed objects   Software modules that are designed to work together but are dispersed throughout the enterprise—a program in one machine sends a message to an object in a remote machine to perform some processing, and the results are sent back to the calling machine.

Document Object Model (DOM)   A standard that allows the manipulation of XML data to access and update the content, structure, and style of documents; it has been referred to as W3C's answer to the Dynamic HTML proposals it has received.

Domain Name Service   A service that allows users to locate computers on the Internet by domain name by translating domain names into the corresponding network address or in the case of TCP/IP, the IP address.

Dynamic HTML   Software that provides the ability to change a Web page's format and layout without downloading an entirely new file.

Dynamic Invocation Interface (DII)   A component of CORBA that allows CORBA objects to find out at run time which objects they may use (Object Management Group).

Dynamic Link Library (DLL)   An executable code module for Microsoft Windows that can be loaded on demand, linked at run time, and then unloaded when the code is no longer needed.

Dynamic personal assistant   Another term for electronic agent; a software program that is autonomous and active; the goal or task is set by the creator/developer.

E-commerce   See Electronic commerce.

EDI (electronic data interchange)   Transmission of data for standard business transactions from one firm's computer to another firm's computer.

EDI boutique   A company that offers EDI translation software and a network via its own Web site; a newer version of an EDI service bureau.

EDMS   See Electronic document management system.

Efficient consumer response   The idea that if all members of the supply chain work together, the end result will be greater customer satisfaction and responsiveness, as well as reduced costs in the form of reduced waste and inventories.

Electronic commerce   Using the Internet, extranets, and intranets to conduct business between trading partners (EDI) and between sellers and customers (electronic shopping).

Electronic data interchange   See EDI.

Electronic document management system (EDMS)   A collection of complementary technologies that bring together document storage, workflow, and indexing.

electronic version of the wallet concept that
…bers, shipping information, and/or digital

…t technology, an object hides (encapsulates) its
…cessed by a well-defined interface that allows
…n any of the operations associated with it.

— A way to transform data into some unread-
…acy (see Decryption).

…guage (XML) A proposed replacement for
…cessed by different programs, delivered by dif-
…wed differently by different users.

…rchitecture based on Internet standards that
…s into the internal network.

…bing made of fiber which is very pure glass,
…ls travel in the form of modulated light; signals
…n over copper wire, but they are able to carry
…n.

…tp) An Internet tool for accessing file archives
…re linked to the Internet.

…of push technology in which only information
…els) that a user specifies is pushed to the user.

…ween networks (often between an internal net-
…that can be designed to keep particular services
…g or leaving; a network security measure.

…oint where an internal network connects to
…rk or to the Internet.

…ased software that allows a group of users to
…ect by providing facilities for communication,
…ct tracking.

…age that browsers see when attaching to a Web
…ins a table of contents (of hypertext links) to
…he site.

…arkup Language) A text-processing markup
…hypertext applications and structure Web-page

…nsport Protocol) The protocol used to move
documents around the Internet.

HTTP cookie See Web cookie.

Hypertext   A link to another file, Web page, or process that is activated by a user clicking on a "hot" area of the screen; the cursor usually changes shape when placed over a "hot" area.

Hypertext Transfer Protocol   See HTTP.

IIOP   See Internet Inter-ORB Protocol.

IIS   See Internet Information Server.

IMAP (Internet Message Access Protocol)   Method of accessing E-mail as if the messages were local.

i-net   The author's term to denote architectures that are based on Internet and Web standards; it includes Internet, intranet, and extranet architectures.

Interface definition language   A language used to describe the interface to a routine or function.

Internet   A network of connected servers that virtually connects the world.

Internet Engineering Task Force   One of two technical working bodies of the Internet Activities Board; the group that focuses on the hardware issues of the Internet, such as TCP/IP and IP addressing.

Internet Information Server (IIS)   A Windows NT–based Web server (Microsoft).

Internet Inter-ORB Protocol (IIOP)   The CORBA message protocol used on the Internet.

Intranet   An internal architecture based on Internet standards.

IP address (Internetwork Protocol address)   The physical address of a computer attached to a TCP/IP network; each node, whether client or server, must have a unique IP address; addresses can be permanently assigned or, for dial-up connections, dynamically assigned for a session.

IP Security protocol (IPSec)   A protocol that enables multiple firewalls from different vendors to establish a secure Internet channel; it includes facilities for encryption, authentication, and key management.

IPX   A NetWare communications protocol used to route messages from one node to another.

IPX-to-IP gateways   A gateway service that converts all IPX requests to IP requests, eliminating the need to put IP connections on each desktop.

Java   A programming language designed primarily for writing Web-based software which is downloaded over the Internet to a client

Electronic wallet   An electronic version of the wallet concept that holds credit-card numbers, shipping information, and/or digital money.

Encapsulation   In object technology, an object hides (encapsulates) its data and code but is accessed by a well-defined interface that allows other objects to perform any of the operations associated with it.

Encryption technology   A way to transform data into some unreadable form to ensure privacy (see Decryption).

Extensible markup language (XML)   A proposed replacement for HTML that can be processed by different programs, delivered by different methods, and viewed differently by different users.

Extranet   An internal architecture based on Internet standards that allows selected outsiders into the internal network.

Fiber-optic cabling   Cabing made of fiber which is very pure glass, over which digital signals travel in the form of modulated light; signals do not move faster than over copper wire, but they are able to carry much more information.

File Transfer Protocol (ftp)   An Internet tool for accessing file archives around the world that are linked to the Internet.

Filtered push   A form of push technology in which only information in the categories (channels) that a user specifies is pushed to the user.

Firewall   A gateway between networks (often between an internal network and the Internet) that can be designed to keep particular services and users from entering or leaving; a network security measure.

Gateway point   The point where an internal network connects to another internal network or to the Internet.

Groupware   Network-based software that allows a group of users to work on a related project by providing facilities for communication, coordination, and project tracking.

Home page   The first page that browsers see when attaching to a Web site—it typically contains a table of contents (of hypertext links) to more information on the site.

HTML (Hypertext Markup Language)   A text-processing markup language used to build hypertext applications and structure Web-page content.

HTTP (Hypertext Transport Protocol)   The protocol used to move documents around the Internet.

HTTP cookie   See Web cookie.

Hypertext   A link to another file, Web page, or process that is activated by a user clicking on a "hot" area of the screen; the cursor usually changes shape when placed over a "hot" area.

Hypertext Transfer Protocol   See HTTP.

IIOP   See Internet Inter-ORB Protocol.

IIS   See Internet Information Server.

IMAP (Internet Message Access Protocol)   Method of accessing E-mail as if the messages were local.

i-net   The author's term to denote architectures that are based on Internet and Web standards; it includes Internet, intranet, and extranet architectures.

Interface definition language   A language used to describe the interface to a routine or function.

Internet   A network of connected servers that virtually connects the world.

Internet Engineering Task Force   One of two technical working bodies of the Internet Activities Board; the group that focuses on the hardware issues of the Internet, such as TCP/IP and IP addressing.

Internet Information Server (IIS)   A Windows NT–based Web server (Microsoft).

Internet Inter-ORB Protocol (IIOP)   The CORBA message protocol used on the Internet.

Intranet   An internal architecture based on Internet standards.

IP address (Internetwork Protocol address)   The physical address of a computer attached to a TCP/IP network; each node, whether client or server, must have a unique IP address; addresses can be permanently assigned or, for dial-up connections, dynamically assigned for a session.

IP Security protocol (IPSec)   A protocol that enables multiple firewalls from different vendors to establish a secure Internet channel; it includes facilities for encryption, authentication, and key management.

IPX   A NetWare communications protocol used to route messages from one node to another.

IPX-to-IP gateways   A gateway service that converts all IPX requests to IP requests, eliminating the need to put IP connections on each desktop.

Java   A programming language designed primarily for writing Web-based software which is downloaded over the Internet to a client machine; HotJava, which is installed on a Web browser, enables the downloaded Java programs to run (JavaSoft).

Java Database Connectivity (JDBC)   An API that allows applets to connect to database servers; based on the X/Open SQL call-level interface (JavaSoft).

Java VM (Java Virtual Machine)   A Java interpreter that converts the Java intermediate language (byte code) into machine language one line at a time and then executes it (JavaSoft—Microsoft also calls its Java Virtual Machine Java VM).

Joint Electronic Payment Initiative (JEPI)   A security protocol that works with SET to handle different payment methods.

Key escrow   Storage of encryption key values in case they must be retrieved by someone other than their original owner.

Lightweight Directory Access Protocol (LDAP)   An Internet-based solution to the directory access protocol which is used to gain access to an X.500 directory listing that permits applications such as E-mail to access directory information through central servers situated at strategic points throughout the network.

Mathematical markup language (MathXL)   An extension of Extensible markup language that addresses the structure and control of mathematical expressions.

Meta-content framework   An extension of Extensible markup language that could be used to design visualization controls, push features, etc.

Multicast   A push architecture that uses a one-to-many output stream—all hosts wishing to receive a particular session are assigned the same IP address, so all subscribers receive a single, shared data stream.

NetPC   Combined network computer and slimmed-down Windows/Intel PC designed for standardized environments.

Network computer   Thin client hardware that has no floppy or hard disk storage, boots from the network, and includes a Web browser and a Java VM for running Java applications.

Network News Transfer Protocol (NNTP)   An extension of the TCP/IP protocol that provides a network news transport service; the standard for Internet exchange of Usenet messages.

Network protocol   A communications protocol used by the network.

Object Linking and Embedding (OLE)   An object-based software technology that allows Windows programs to exchange information and work together (Microsoft).

Object request broker   A technology that takes a message from a client program, locates the target object class, finds the object occurrence, performs the necessary translation, and passes back the result.

Objects Independent program modules designed to work together at run time without prior knowledge of one another.

ODBC (Open Database Connectivity) An API that provides access to relational databases (Microsoft).

OLAP (on-line analytic processing) Structuring data in multidimensional cubes to facilitate analysis.

OLE See Object Linking and Embedding.

On-line analytic processing See OLAP.

On-line transaction processing Transaction processing that happens in real time.

Open Buying on the Internet (OBI) An architecture for real-time, mass-volume transactions between businesses and their suppliers.

Open Database Connectivity See ODBC.

Open Financial Exchange (OFX) An extension of Extensible markup language that will handle payment instructions to banks.

Open User Recommended Solutions (OURS) A nonprofit industry group dedicated to promoting open IT standards that can be used to link intranets to specific business functions.

OpenDoc Similar to OLE; designed as a cross-platform technology that is platform-independent and interoperates with other similar architectures such as OLE.

ORB See Object request broker.

Packet filtering Examining the source and destination addresses and ports of incoming traffic and denying or allowing packets to enter based on a set of rules; a type of firewall.

Password cracking An intruder stealing a password file and using the passwords to gain access.

Password sniffing An intruder capturing passwords to gain access.

Persistent A persistent connection is one which tries to connect to the same resources the next time a user logs in. The keepalive function of HTTP 1.1 is an example of a persistent connection—it keeps a route or "tunnel" open all the way through the network, allows persistent connections, and remains open for multiple requests.

Ping o' death An intruder sending a "ping" message that crashes a system.

Plug-in   Code that performs one function for specific file types and is platform-specific; executes on the desktop while the user is in the browser; downloads from the Web server.

POP3 (Post Office Protocol version 3)   An E-mail server protocol used in the Internet; POP is used to get the mail, and SMTP (Simple Mail Transfer Protocol) is used to download it; being replaced by IMAP (Internet Message Access Protocol).

Private-key encryption   Senders and receivers use the same key to encrypt and decrypt messages.

Protocol   Rules and formats for communications across different devices.

Prototyping   Developing a working model of an application during the design stage.

Proxy server   Restricts Web access to selected people by preventing unauthorized inbound traffic and restricting downloading; a type of firewall.

Public-key encryption   Senders and receivers use both a common key (public key) and a private key to encrypt and decrypt messages.

Publish and subscribe   Push technology used internally when a company combines internal data with external data from content providers into categories to which internal users subscribe.

Push technology   Technology that automatically sends a user information rather than waiting for the user to ask for it.

Quality of service   The ability to define a level of performance in a system.

Rapid application development   A development methodology that delivers working pieces of an application incrementally.

Remote maintenance   Automated management services, including virus scans, software updates, and ongoing maintenance, performed on client machines during off-hours.

Remote Method Invocation (RMI)   A technology that enables Java applications to communicate over a lightweight protocol.

Remote procedure call   See RPC.

Resource Reservation Protocol   A new protocol for the Internet that delivers traffic based on the required level of service, so that certain traffic, such as videoconferences, will be delivered before other traffic, such as E-mail.

RPC (remote procedure call)   A client process calls a function on a remote server and waits for the results before continuing with its processing.

Secure Electronic Transactions   See SET.

Secure Hypertext Transfer Protocol   See S-HTTP.

Secure Multipart Internet Mail Encoding   See S/MIME.

Secure Socket Layer   See SSL.

Security protocol   A communications protocol that encrypts and decrypts a message for on-line transmission.

Self-defining   Contents of executable components are determined at execution time.

SET (Secure Electronic Transactions)   A specification for on-line commerce that uses public-key-encrypted digital signatures.

SET Mark   The label displayed on sites that use SET-certified commerce software.

SGML (Standard Generalized Markup Language)   An ISO standard for defining the formatting in a text document which became popular as a result of the tremendous increase in electronic publishing; HTML, the format language for Web pages, is a subset of SGML.

S-HTTP (Secure Hypertext Transfer Protocol)   An extension of HTTP for authentication and data encryption between a Web server and a Web browser.

Simple Network Management Protocol   See SNMP.

Smart card   A credit-card-like device with an embedded microchip that requires a special reader.

S/MIME (Secure Multipart Internet Mail Encoding)   MIME is a protocol for transmitting mixed-media files across TCP/IP networks. S/MIME is a more secure version of MIME.

SNMP (Simple Network Management Protocol)   A common format used by network devices to exchange management information with the network management console(s).

Spamming   Sending electronic junk mail that brings a network's performance to a standstill.

Spoofing   When a hacker poses as a legitimate host using a fabricated IP address.

SSL (Secure Socket Layer)   A protocol that secures communication channels between clients and servers, and encrypts transmissions (Netscape Communications).

Standard Generalized Markup Language   See SGML.

Stateful inspection   Inspection of entire packets, comparing them to known bit patterns of friendly packets; a type of firewall.

Stateless   Does not maintain information across sessions.

Storefront   An electronic store accessible via the Internet that offers items for sale and is capable of handling the financial transaction on-line.

Symmetric encryption   A data encryption method that uses the DES standard, which uses the same key on both ends of a transmission.

SYN Flood   An intruder flooding a site with E-mail messages.

T1   A network channel that can handle 24 voice or data channels at 64 kbps, giving a capacity of 1.544 Mbps.

T3   A channel that combines 28 T1 lines, giving a capacity of 44 Mbps; requires fiber-optic cabling.

TCP/IP (Transmission Control Protocol/Internetwork Protocol) A de facto standard for interconnecting otherwise incompatible computers; the connection method used by the Internet.

Thin client   A client machine that performs very little data or application processing—the processing is done in the server, and the client machine processes the output for the screen display.

Threaded discussion   A running log of electronically generated comments about a subject; used in chat rooms of on-line services and in groupware.

Token   A small credit-card-size device that a remote user carries to provide authentication.

Transaction-processing monitors (TP monitors)   Devices that manage transactions across multiple servers.

Tunneling   Encapsulating packets of data written for one network protocol inside packets used for another.

Unicast   A one-to-one push architecture—each subscriber is sent an identical data stream.

URL (uniform resource locator)   The name given to a site in the Internet which is used to access the site.

User Datagram Protocol   A TCP/IP protocol describing how messages reach application programs within a destination computer; normally bundled with IP-layer software.

Virtual private network   Use of encryption technology to ensure the secure passage of messages over the Internet to employees and strategic trading partners such as branch offices and suppliers.

Web   Another name for the World Wide Web.

Web browser   Software that allows a user to navigate the interconnected documents on the World Wide Web.

Web cookie   Text files of information sent back and forth between a Web site and a Web browser.

Web hosting   Internet service providers offering to put up Web sites for outside companies on computers owned by the providers and charging fees based on equipment and transmission capacity used.

Web page   A page in a WWW-based document or application.

Web server   A computer that contains the documents and files that are displayed when users access the server via HTTP.

Web site   A set of Web pages that can be visited by Web browsers.

Webcasting   A term originally applied to push technology, but coming to refer to the process of sending live video programming to several Internet users simultaneously.

Webmaster   The person(s) responsible for the management and often the design of a company's Web site; a female might be called a Web-mistress.

Windows DNA   A technology that combines the features from the Internet and client/server architectures into a framework for an application to run over the Web on any platform.

X.12   ANSI standard for EDI that is designed to support cross-industry exchange of business transactions; it specifies the vocabulary (dictionary) and format for electronic business transactions.

X.509   A specification for a certificate which binds an entity's distinguished name to its public key through the use of a digital signature; also contains the distinguished name of the certificate issuer.

Zero-Administration for Windows (ZAW)   The ability to manage client machines from a single location; it includes server software that runs on Windows NT and communicates with client software running on your Windows 95- and 98-based systems.

ACRONYMS

3GL	Third-generation language
4GL	Fourth-generation language
ANSI	American National Standards Institute
API	Application programming interface
ARIN	American Registry for Internet Numbers
ASCII	American Standard Code for Information Interchange
ASP	Active Server Page (Microsoft)
ATM	Asynchronous transfer mode
BOOTP	Bootstrap Protocol
CDF	Channel Definition Format (Microsoft)
CERT	Computer Emergency Response Team
CGI	Common gateway interface
CICS	Customer Information Control System (IBM)
CIDR	Classless Interdomain Routing
CIFS	Common Internet File System
CIM	Common Information Model
COM	Component Object Model (Microsoft)
COO	Cost of ownership
CORBA	Common Object Request Broker Architecture (Object Management Group)
DBMS	Database management system
DCOM	Distributed Component Object Model (Microsoft)
DHC	Dynamic Host Configuration Group of IETF
DHCP	Dynamic Host Configuration Protocol
DII	Dynamic Invocation Interface (Microsoft)
DLL	Dynamic Link Library (Microsoft)
DMA	Document Management Alliance
DMI	Desktop Management Interface
DMTF	Desktop Management Task Force
DNS	Domain Name Service
DOM	Document object model
DSS	Decision support system
ECR	Efficient Consumer Response
EDI	Electronic data interchange
EDIFACT	Electronic Data Interchange for Administration Commerce and Transport
EDMS	Electronic document management system
ftp	File transfer protocol
GEIS	General Electric Information Services
GIOP	General Inter-ORB Protocol (CORBA—Object Management Group)
GUI	Graphical user interface
HMMA	Hypermedia Management Application (WBEM)
HMMO	Hypermedia Managed Object (WBEM)

HMMP	Hypermedia Management Protocol (WBEM)	Kbytes	One thousand (kilo) bytes
		LAN	Local area network
HMMS	Hypermedia Management Schema (WBEM)	LDAP	Lightweight Directory Access Protocol
HMOM	Hypermedia Object Manager (WBEM)	MAPI	Messaging API
HTML	Hypertext Markup Language	MathML	Mathematical Markup Language
HTTP	Hypertext Transport Protocol	Mbps	One million bits per second
I/O	Input/output	Mbytes	One million bytes
ICAPI	Internet Connection API (IBM)	MCF	Meta Content Framework
IDC	International Data Corp.	MHz	Megahertz
IDL	Interface definition language	MTS	Microsoft Transaction Server
IETF	Internet Engineering Task Force	NC	Network computer
IIOP	Internet Inter-ORB Protocol (CORBA—Object Management Group)	NDS	Novell Directory Services
		NetBIOS	Network Basic Input/Output Operating System
IIS	Internet Information Server (Microsoft)	NetPC	Network PC
		NFS	Network File System (Sun)
IMAP4	Internet Message Access Protocol version 4	NNTP	Network News Transport Protocol
		NPV	Net present value
InterNIC	Internet Network Information Center	NSAPI	Netscape Server API
IP	Internetwork Protocol	OBI	Open Buying on the Internet
IPOC	Interim Policy Oversight Committee	ODBC	Open Database Connectivity (Microsoft)
IPSec	IP Security Protocol	ODMA	Open Document Management API
IPv4	IP addressing version 4	OFC	Open Financial Connectivity (Microsoft)
IPv6	IP addressing version 6		
IRC	Internet Relay Chat	OFX	Open Financial Exchange (Microsoft)
ISAPI	Internet Services API (Microsoft)	OLAP	On-line analytical processing
ISP	Internet service provider	OLE	Object Linking and Embedding (Microsoft)
IT	Information technology		
JDBC	Java Database Connectivity (JavaSoft)	OLTP	On-line transaction processing
JEPI	Joint Electronic Payment Initiative	ORB	Object Request Broker
JIT	Just-in-time	OURS	Open User Recommended Solutions
JMAPI	Java Management API (JavaSoft)	PC	Personal computer
Kbps	One thousand (kilo) bits per second		

PCT	Private Communications Technology
PHF	Packet Handling Function (file)
POP3	Post Office Protocol version 3
QOS	Quality of service
RAM	Random-access memory
RARP	Reverse Address Resolution Protocol
RFC	Request for Comment (IETF)
RMI	Remote Method Invocation (Sun)
ROI	Return on investment
RPC	Remote procedure call
RSVP	Resource Reservation Protocol
S/MIME	Secure/Multipurpose Internet Mail Extensions
SET	Secure Electronic Transactions
SGML	Standard Generalized Markup Language
S-HTTP	Secure Hypertext Transfer Protocol
SKU	Stock-keeping unit
SMB	Server Message Block (Microsoft)
SMTP	Simple Mail Transfer Protocol
SNMP	Simple Network Management Protocol
SOM	System Object Model (IBM)
SQL	Standard Query Language
SSL	Secure Socket Layer (Netscape)
TCP/IP	Transmission Control Protocol/Inter-network Protocol
TLS	Transport layer security
TP	Transaction processing
UDP	User Datagram Protocol
URL	Uniform resource locator
VAR	Value-added reseller
VPN	Virtual private network
W3C	World Wide Web Consortium
WAN	Wide area network
WBEM	Web-based Enterprise Management
WfMC	Workflow Management Coalition
WWW	World Wide Web
XML	Extensible Markup Language
ZAW	Zero Administration Windows

INDEX

About the Author

Dawna Travis Dewire, who has nearly 30 years of computer-related experience, is the author of the bestselling *Client/Server Computing* and *Second Generation Client/Server Computing*. A private consultant who lives in Wellesley, Massachusetts, she teaches at Babson College.